CONCRETE TECHNOLOGY

CONCRETE TECHNOLOGY

Materials Science of Concrete Special Volume

PROCEEDINGS OF THE ANNA MARIA WORKSHOPS
2002: Designing Concrete for Durability
2003: Testing and Standards for Concrete Durability
2004: Cement and Concrete of the Future

Editors
Jan Skalny
Sidney Mindess
Andrew Boyd

Published by

The American Ceramic Society
735 Ceramic Place
Suite 100
Westerville, Ohio 43081
www.ceramics.org

Concrete Technology

For information on ordering titles published by The American Ceramic Society, or to request a publications catalog, please call 614-794-5890, or visit www.ceramics.org.

ISBN 1-57498-268-0

10 09 08 07 06 5 4 3 2 1

Contents

Preface

The Fifth Anna Maria Workshop was held on November 17-19, 2004 at the City Hall of Holmes Beach, Florida. The topic of the meeting was The Future of the Cement and Concrete Industries, an issue close to the hearts of all the participants. The Workshop was attended by about 50 people representing industry, academia, and professional organizations, and its deliberations focused on the successes and future challenges of the cementitious construction materials industry.

In addition to the 2004 Anna Maria Workshop, this proceedings includes selected papers and other unpublished documents from the previous Anna Maria Workshops that covered the topics of Designing Concrete for Durability (2002) and Testing and Standards for Concrete Durability (2003). The underlying concern of all three workshops was the issue of preservation and improvement of service life of the concrete-based infrastructure under changing economic, business and technical conditions. Together, they present a more complete picture of the general philosophy of the Anna Maria Workshops.

The members of the Organization Committee would like to express their thanks to all participants for their active involvement in the workshops' deliberations, the authors of the presentations for their timely and high quality efforts, and the reviewers of the manuscripts. Special thanks are due to chairmen of the focus groups and a few other individuals who were instrumental in preparing the 2003 Group Reports.

We all are most thankful to the sponsors of the workshops; without their moral and financial support, these workshops would not be possible. We especially appreciate that numerous high-level executives of the sponsoring organizations have found the time to attend the workshops.

Special thanks are due to the University of Florida, Department of Civil and Coastal Engineering in Gainsville, Florida – and specifically its Chairman, Professor Joseph Tadesco – for sponsoring the Workshop organization. Thanks are due also to the students who were responsible for the hour-by-hour smooth progress of the workshop and social activities.

The continuous support by Holmes Beach Mayor Carol Whitmore and her staff is respectfully acknowledged. It is thanks to their dedicated efforts that the beautiful environment of the Anna Maria Island attracts tourists and professional activities.

Finally, we would like to thank The American Ceramic Society staff for professional help and dedication in making publication of these proceedings possible.

May 2005

Jan Skalny
Sidney Mindess
Andrew Boyd

ANNA MARIA WORKSHOP 2004
CEMENT AND CONCRETE OF THE FUTURE
Anna Maria Island, Florida - November 17-19, 2004

Arnon Bentur	Technion / W.R. Grace	Israel / USA
Andrew Boyd	University of Florida	USA
Chuck Cornman	W.R. Grace	USA
Tate Coverdale	Master Builders Technologies	USA
Anik Delagrave	Lafarge Canada	Canada
Sidney Diamond	Purdue University	USA
Peter Emmons	Structural Group	USA
Emery Farkas	Consultant	USA
Stephen Forster	Federal Highway Administration	USA
Juraj Gebauer	Holcim Group Support	Switzerland
Alan Gee	Heidelberg Cement	USA
Frederick Glasser	University of Aberdeen	UK
Duncan Herfort	Aalborg Portland	Denmark
Doug Hooton	University of Toronto	Canada
Charles Ishee	Florida Department of Transportation	USA
Lesley Suz-Chung Ko	Holcim Group Support	Switzerland
Richard Lee	R J Lee Group	USA
Waltter Lopez-Gonzales	CEMEX	USA
Jacques Lukasik	Lafarge	France
Marjorie Lynch	Surtreat NE	USA
Jacques Marchand	Laval University	Canada
Denis Mitchell	McGill University	Canada
Jeff O'Leary	Florida Rock	USA
Jan Olek	Purdue University	USA
Ivan Petrovic	Englehard Corporation	USA
Jim Pierce	Bureau of Reclamation	USA
Richard Reaves	Troxler Electronic Laboratories	USA
David Rosenberg	Hycrete	USA
Karen Scrivener	Swiss Federal Institute of Technology B Lausanne	Switzerland
Jan Skalny	Materials Service Life	USA
David Stokes	FMC Lithium Division	USA
Ed Sullivan	Portland Cement Association	USA
Bill Suojanen	Suojanen Law	USA
Larry Sutter	Michigan Technical University	USA
Peter Taylor	CTL	USA
Niels Thaulow	R J Lee Group	USA
Paul Tourney	Materials Service Life	USA
Markus Tschudin	Holcim Group Support	Switzerland
Danica Turk	Consultant	USA

ANNA MARIA WORKSHOP 2003
CONCRETE DURABILITY: TESTING & STANDARDS
Anna Maria Island, Florida - November 5-7, 2003

Emmanuel Attiogbe	Master Builders Technologies	USA
Arnon Bentur	Technion / W.R. Grace	Israel / USA
Andrew Boyd	University of Florida	USA
Andrzej Brandt	Polish Academy of Science	Poland
Nick Buenfeld	Imperial College	UK
Neil Cumming	Levelton Engineering	Canada
Sharon DeHayes	Rinker Materials	USA
Sidney Diamond	Purdue University	USA
Kevin Folliard	University of Texas @ Austin	USA
Stephen Forster	Federal Highway Administration	USA
Geoff Frohnsdorff	National Institute of Standards & Technology (NIST)	USA
Juraj Gebauer	Holcim Group Support	Switzerland
Alan Gee	Heidelberg Cement	USA
Carolyn Hansson	University of Waterloo	Canada
Nataliya Hearn	University of Windsor	Canada
Doug Hooton	University of Toronto	Canada
Charles Ishee	Florida Department of Transportation	USA
Lesley Suz-Chung Ko	Holcim Group Support	Switzerland
Richard Lee	R J Lee Group	USA
Jacques Lukasik	Lafarge	France
Marjorie Lynch	Surtreat NE	USA
Jacques Marchand	Laval University	Canada
Donald Meinheit	Wiss, Janney, Elstner Associates	USA
Sidney Mindess	University of British Columbia	Canada
Denis Mitchell	McGill University	Canada
David Myers	W.R. Grace	USA
Jeff O'Leary	Florida Rock	USA
Jan Olek	Purdue University	USA
Jim Pierce	Bureau of Reclamation	USA
Ken Rear	Heidelberg Cement	USA
Richard Reaves	Troxler Electronic Laboratories	USA
J.C. Roumain	Holcim	USA
Jan Skalny	Materials Service Life	USA
Ken Snyder	National Institute of Standards & Technology (NIST)	USA
Larry Sutter	Michigan Technical University	USA
Peter Taylor	CTL	USA
Joseph Tedesco	University of Florida	USA
Niels Thaulow	R J Lee Group	USA
Albert Tien	Holcim Group Support	Switzerland
Paul Tourney	Materials Service Life	USA
Danica Turk	Consultant	USA
Suneel Vanikar	FHWA	USA
Carl Walker	CW Consulting	USA

ANNA MARIA WORKSHOP 2002
DESIGNING CONCRETE FOR DURABILITY
Anna Maria Island, Florida - November 13-15, 2002

Carmen Andrade	Instituto Eduardo Torroja – CSIC	Spain
Arnon Bentur	Technion – Israel Institute of Technology	Israel
Andrew Boyd	University of Florida	USA
Paul Brown	Pennsylvania State University	USA
Norah Crammond	Building Research Establishment	UK
David Darwin	University of Kansas	USA
Sharon DeHayes	Rinker Materials	USA
Anik Delagrave	Lafarge Canada	Canada
Sidney Diamond	Purdue University	USA
Christos Drakos	University of Florida	USA
Emery Farkas	Consultant	USA
Kevin Folliard	University of Texas - Austin	USA
Stephen Forster	Federal Highway Administration	USA
Geoff Frohnsdorff	National Institute of Standards & Technology (NIST)	USA
Juraj Gebauer	Holcim Group Support	Switzerland
Alan Gee	Heidelberg Cement Group	USA
Carolyn Hansson	University of Waterloo	Canada
Doug Hooton	University of Toronto	Canada
Charles Ishee	Florida Department of Transportation	USA
Felek Jachimowicz	W.R. Grace	USA
Ulla Jakobsen	Concrete Experts International	Denmark
Fred Kinney	ESSROC Materials	USA
Richard Lee	R J Lee Group	USA
Jacques Lukasik	Lafarge	France
Jacques Marchand	Laval University	Canada
Tom McCall	Tamms Industries Inc.	USA
Matt Miltenberger	Master Builders Technologies	USA
Sidney Mindess	University of British Columbia	Canada
Denis Mitchell	McGill University	Canada
Denis Montgomery	University of Wollongong	Australia
Jan Olek	Purdue University	USA
Jim Pierce	Bureau of Reclamation	USA
Ken Rear	Heidelberg Cement Group	USA
Jean-Claude Roumain	Holcim	USA
Karen Scrivener	Swiss Federal Institute of Technology – Lausanne	Switzerland
Jan Skalny	Materials Service Life	USA
Lesley Ko Suz-Chung	Holcim Group Support	Switzerland
Larry Sutter	Michigan Technical University	USA
Joseph Tedesco	University of Florida	USA
Niels Thaulow	R J Lee Group	USA
Paul Tourney	Materials Service Life	USA
David Trejo	Texas A&M University	USA
Danica Turk	Consultant	USA
Michelle Wilson	Portland Cement Association	USA

Thanks are due to the following organizations that supported the 2004 Workshop:

CEMEX

Degussa Admixtures

Florida Rock Industries

Grace Construction Products

Heidelberg / Lehigh Cement

Holcim Group Support - Switzerland

HOLCIM - US

Lafarge Corporation

Materials Service Life

Portland Cement Association

Rinker Materials

RJ Lee Group

University of Florida

I. Papers Presented at the 3rd Anna Maria Workshop (2002)

COMPRESSIVE STRENGTH: THE WRONG WAY TO ASSESS CONCRETE

Sidney Mindess
University of British Columbia
Vancouver, Canada

> *"Nearly all the scientific principles which constitute the foundation of civil engineering are susceptible of complete and satisfactory explanation to any person who* really *possesses only so much elementary knowledge of arithmetic and natural philosophy as is* supposed *to be taught to boys of twelve or fourteen in our public schools.*
>
> *.........one of the leading objects has been to elucidate in plain English, a few important elementary principles which the savants have enveloped in such a haze of mystery as to render pursuit hopeless to any but a confirmed mathematician.*
>
> *- J. C. Trautwine*[1]

ABSTRACT It is argued that the standard tests of compressive strength (or of splitting tensile strength) are often an inappropriate means of assessing the quality of *in situ* concrete. The compressive strength is not a fundamental concrete property, since its value is largely a function of just how it is measured. It is generally accepted that, under almost any type of loading, failure occurs in tension. It is thus suggested that a direct measure of tensile strength would be the best indicator of concrete quality, and particularly of concrete damage.

1. INTRODUCTION

In modern building codes, such as the Uniform Building Code[2] or ACI 318, Building Code Requirements for Reinforced Concrete[3], the only criterion for judging the adequacy of concrete in a structure is the compressive strength, f'_c. For new construction, the acceptance criterion for concrete is that the required average compressive strength, f'_{cr}, exceed f'_c by a factor which takes into account the standard deviation of the test data.

For concrete *in situ* which is suspected of having undergone damage, the acceptance criterion[2,3] is that the compressive strength measured on drilled cores shall be $\geq 0.85 f'_c$.

Implicit in these criteria is the attempt to balance the risk that "bad" concrete will be accepted against the risk that "good" concrete will be rejected. The question to be explored below is: Are these criteria sufficient?

2. WHAT IS f'_c ?

In North America, the basic *cement* tests were standardized in 1900. However, there were by then still no generally agreed upon procedures for testing concrete strength. Indeed, there was still some controversy as to whether compressive strength or tensile strength best characterized the mechanical properties of concrete. For instance, the thirteenth edition of Trautwine's[1] *The Civil Engineer's Pocket-Book* (1888), in its brief discussion of concrete construction, does not mention concrete testing. Sabin's[4] *Cement and Concrete* (1905) describes a variety of compressive tests, mostly carried out on cubes of up to 12 inches on a side, sometimes with the loading platens having smaller lateral dimensions than the cubes themselves. But no "standard" test method was advocated, and Sabin appeared not to consider this to be important. As well, in discussing cement tests, he stated that

> "Although tests of compressive strength are of interest from a scientific point of view, it is not considered that they would give much greater information concerning the relative qualities of cements than is given by tensile tests, and therefore they need not be included in an ordinary series of acceptance tests."

And, foreshadowing later studies on the meaning of the compressive strength, Sabin went on to say

> "In practically all forms of masonry construction, cement is called upon to resist compression. In consequence of this fact, the opinion is somewhat general that the greatest amount of information would be obtained by compression tests. *But the compressive strength of cement is so much greater than its tensile strength, that when failures occur, they are likely to be due to other forms of stress.*" (Italics mine).

The most common North American test standard, ASTM C 39[5], *Standard Test Method for Compressive Strength of Cylindrical Concrete Specimens*, did not appear until 1921, and was then still designated as a "tentative" standard. (The cube tests, which are widely used in other parts of the world, were standardized at about the same time). However, it soon became apparent that this test had severe limitations, being highly dependent on the precise testing conditions. Indeed, even the phrase "compressive strength" can be rather misleading, since concrete does not fail due to the compressive stresses, but due to the secondary tensile stresses which are induced by compressive loading.

As early as 1927, Brandtzaeg[6] argued that splitting or cleavage failure will occur when the lateral tensile stress or strain reaches a limiting value across the plane parallel to the direction of maximum compression. Subsequently, as part of one of the first detailed studies of cracking and fracture of concrete, Richart, Brandtzaeg and Brown[7] concluded that under uniaxial compression, concrete specimens fail as a result of lateral (tensile) splitting. Later researchers have reached much the same conclusions. For instance, the idealized stresses around a single aggregate particle in concrete under uniaxial compression were described by Vile[8] as shown in Fig. 1.

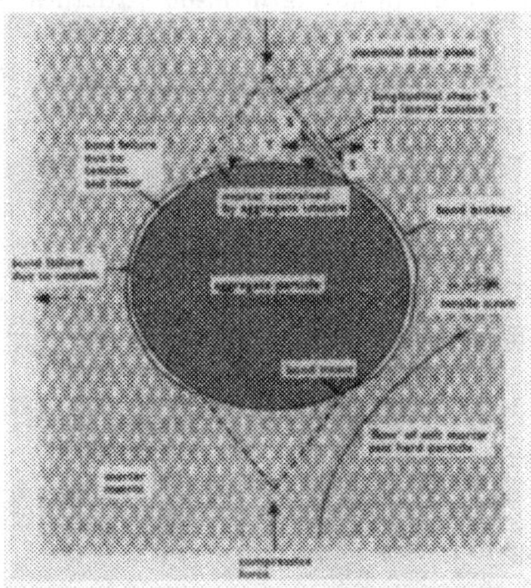

Fig. 1. Idealization of stresses around a single aggregate particle at the discontinuity point under uniaxial compression

Stress analysis indicated the following order of failure: (i) tensile bond failure; (ii) shear bond failure; (iii) tensile matrix failure; and (iv) occasional aggregate failure. That is, under compression, there are some zones around an aggregate particle subjected to tensile stress. When the stress reaches the concrete tensile strength, a crack will begin to propagate. Failure is eventually due primarily to the lateral tensile stresses produced by the microstructural failure of the material. More recently, Neville[9] described the failure process in compression by considering the limiting strains:

> "[W]e do not know the exact criteria of failure of concrete but there are strong indications that failure occurs at a limiting strain of 0.002 to 0.004 in compression or 0.0001 to 0.0002 in tension. Because the ration of the latter of these strains to the former is less than Poisson's ratio of concrete, it follows that conditions of failure in circumferential tension are achieved before the limiting compressive strain has been reached."

3. WHY DO WE DETERMINE f'_c?

It is stated clearly in ASTM C 39[5] that "Care must be exercised in the interpretation of the significance of compressive strength determinations by this test method since strength is not a fundamental or intrinsic property of concrete made from given materials." Why, then, do we obsessively measure f'_c not only in construction practice but also in research? There are a number of reasons:

- The test is relatively easy and inexpensive to carry out.
- Values of f'_c are essential for the current strength-based structural design codes.
- We have "always" measured f'_c, at least since 1921; that is, for all practical purposes, since a "Time whereof the memory of man runneth not to the contrary".
- The measurement of f'_c is extremely useful for routine quality control purposes.

However, we must also remember what f'_c is NOT:

- A true measure of the "strength" of concrete
- A general indicator of concrete quality
- A measure of durability
- An indicator of possible damage to *in situ* concrete

4. IS THERE A BETTER ALTERNATIVE TO f'_c?

There is little doubt that the direct tensile strength would be the best indicator of concrete damage, since, as we have seen, concrete almost always fails in tension, regardless of the nature of the applied loading. Thus, a tension test would be much more sensitive to internal cracking than would a compression test, which would tend to close rather than open cracks. However, direct tension tests are rarely carried out, because of the difficulty in carrying out such tests. The secondary stresses induced through gripping make the test results difficult to reproduce or to interpret. Also, there is as yet no generally accepted test method for carrying out direct tension tests. (There is a RILEM recommendation[10] for a direct tension test, which involves gluing end plates on to the specimen, but there are as yet no ASTM test).

Most commonly, the tensile strength is estimated from the splitting tension test (ASTM C 496[11]), which is shown schematically in Fig. 2. While this test is, like the compression test, relatively easy and economical to carry out, it also has some of the same drawbacks. The values obtained depend upon the specimen size and the width and properties of the bearing strips. Indeed, the Uniform Building Code[2] states explicitly that "[s]plitting tensile strength tests shall not be used as a basis for field acceptance of concrete". Moreover, because of the nature of the test geometry, the specimen is forced to fail on a predetermined plane, rather than on its weakest plane. Thus, the tensile strength determined in this way can also not be considered to be a fundamental or intrinsic property of the concrete.

We are, therefore, faced with a serious dilemma: The direct tension test is too difficult and expensive to carry out routinely, and the most common indirect tension test does not provide a true, unambiguous value.

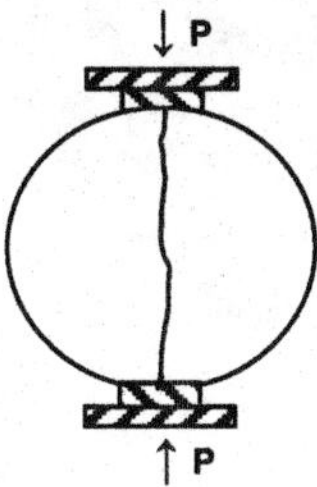

Fig. 2. Schematic of splitting tension test

5. PRESSURE TENSION TEST

One way around the problem may be to adopt a test similar to the so-called "pressure" tension test, which was developed by the British Research Establishment[12, 13] in the 1970's. In this test, which is shown schematically in Fig. 3, gas pressure is applied to the curved surface of a standard concrete cylinder, but not to the ends. This induces a tensile stress on a plane perpendicular to the longitudinal axis of the cylinder, and what looks like a cleavage failure occurs on the weakest plane within the loaded portion of the specimen. This is, in effect, a biaxial stress test, which has been used successfully as an indicator of damage due to alkali-aggregate reaction[14] and to sulphate attack[15]. And, like the standard compression test and splitting tension test, the pressure tension test is also easy and economical to perform.

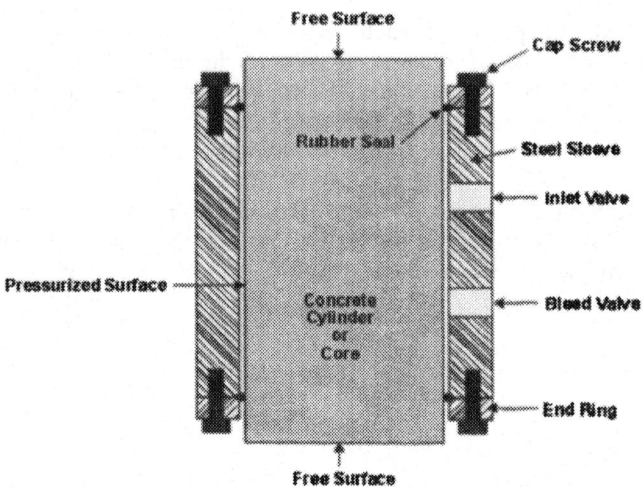

Fig. 3. Schematic of pressure tension device

5. CONCLUSIONS

1. Current design codes are based almost entirely on compressive strength.
2. Concrete fails primarily by tensile cracking, regardless of the type of loading.
3. Neither the standard compression test (ASTM C 39) nor the standard splitting tensile test (ASTM C 496) provides fundamental or intrinsic material properties for concrete.
4. There is, therefore, a need to develop proper procedures for determining the true direct tensile strength of concrete.

REFERENCES

[1]J.C. Trautwine, Jr., *The Civil Engineer's Pocket-Book*, 13th ed., John Wiley & Sons, New York and E. & F.N. Spon, London, 1888.

[2]*Uniform Building Code*, International Conference of Building Officials, Whittier, California, 1997.

[3]ACI Committee 318, *Building Code Requirements for Reinforced Concrete*, American Concrete Institute, Farmington Hills, Michigan, 2002.

[4]L.C.Sabin, *Cement and Concrete*, Archibald Constable and Co., London, 1905.

[5]ASTM C 39, *Standard Test Method for Compressive Strength of Cylindrical Concrete Specimens*, American Society for Testing and Materials, West Conshohocken, PA, 2002.

[6]A, Brandtzaeg, "Failure of a Material Composed of Non-Isotropic Elements," *Kongelige Norske Videnskabernes Selskabs Forhandlinger Skrifter*, No. 2, 1927, pp. 68.

[7]F.F. Richart, A Brandtzaeg and R.L. Brown, "A Study of the Failure of Concrete Under Combined Compressive Stresses," *Bulletin 185*, Engineering Experiment Station, University of Illinois, 1928.

[8]G.W.D. Vile, "The Strength of Concrete under Short-term Static Biaxial Stress"; pp. 275-288 in *The Structure of Concrete*. Edited by A.E. Brooks and K. Newman. Cement and Concrete Association, London, 1968.

[9]A.M. Neville, *Properties of Concrete*, 4th ed., John Wiley & Sons, 1996.

[10]RILEM, CPC7, Direct Tension of Concrete Specimens, pp. 23-24 in *RILEM Technical Recommendation for the Testing and Use of Construction Materials*, E. & F.N. Spon, London, 1994.

[11]ASTM C 496, *Standard Test Method for Splitting Tensile Strength of Cylindrical Concrete Specimens*, American Society for Testing and Materials, West Conshohocken, PA, 2002.

[12]N. Clayton and F.J. Grimer, "The Diphase Concept, with Particular Reference to Concrete," pp. 283-318 in F.D. Lydon, ed., *Developments in Concrete Technology –1,* Applied Science Publishers, London, 1979.

[13]A. Boyd and S. Mindess, "An Indirect Tension Test for Concrete," pp. 590-594 in Vol. I, *Proceedings of the 5th International Symposium on Cement and Concrete*, Shanghai. Tongji University Press, Shanghai, China, 2002.

[14]T.W. Bremner, A.J. Boyd, T.A. Holm and S.R. Boyd, "Indirect Tensile Testing to Evaluate the Effect of Alkali-Aggregate Reaction in Concrete," in *Proceedings, Structural Engineering World Wide 1998*, San Francisco, CA, 1998.

[15]A.J. Boyd and S. Mindess, "The Effects of Sulfate Attack on the Tensile to Compressive Strength Ratio of Concrete," pp. 789-796 in N. Banthia, K. Sakai and O.E. Gjorv, eds., *Proceedings Third International Conference on Concrete under Severe Conditions, CONSEC '01,* Vancouver, University of British Columbia, Vancouver, Canada, 2001.

DURABILITY DESIGN OF CONCRETE COVER *THE KNOWING-DOING GAP*

Arnon Bentur
National Building Research Institute - Faculty of Civil Engineering
Technion, Israel Institute of Technology, Israel

ABSTRACT It is demonstrated that the most efficient engineering tool to control durability performance, in particular for the major issues of steel corrosion and sulfate attack, is the permeability of concrete. The know-how of making impermeable concrete is well established but unfortunately it is not rooted effectively in design and field practice. One of the drawbacks is the limitation of the strength criteria, which is widely accepted as the quality control parameter of concrete, but does not provide adequate estimate of service life. This is demonstrated in this paper in quantitative terms, in particular with respect to curing and its influence on strength and service life. Thus, there is a need to re-educate the design and construction engineers to take all the necessary steps to achieve impermeability and highlight the limitations of the strength criteria when considering durability. One of the ways to bring this message across is to demonstrate the differences in relative effects of curing and strength on durability performance. In order to implement this approach in practice much more effort should be dedicated to the development of tests for evaluating the performance of the concrete cover.

1. INTRODUCTION

Durability of concrete and reinforced concrete has become in recent decades a major issue in civil engineering, reflecting the damage caused by deteriorating structures. This state of affairs has promoted in depth scientific studies of deterioration processes, to resolve their chemical and physical effects as well as their kinetics. Within this context studies and modeling of transport properties have been advanced quite impressively, looking into the range of transport mechanisms and their cumulative influence.

However, this level of fundamental understanding has not diffused into practice, as is evident to some extent in recent reviews [1-3]. These papers highlight the increased severity of the durability issue, which seems to be greater than experienced decades ago. They emphasize the limitations associated with specifying and control of concrete quality on the basis of strength. Within this context attention is usually given to several influences: (i) The higher strength of modern cements enabling the production of stronger concretes at higher w/c ratio, resulting in larger permeabilities, (b) The difference between the concrete cover properties and the core properties which are not reflected in quality control based on strength, and (c) The range of cement compositions with a variety of additions, where the relations between strength and permeability and durability can be markedly different than those of ordinary Portland cement. Also highlighted are influences of site practices such as curing which may have detrimental effect on durability and can not be detected by strength quality control.

Attention is always given in these reviews to the limitations of the current standards which set requirements that are prescriptive in nature, based on specifying maximum w/c ratio, minimum cement content and strength. Even the new European standard EN 206, which is more detailed in the sense that it addresses the type of corrosion process and the environmental conditions is prescriptive in nature (Table 1).

The EN standard, like all other standards, does not address directly the issue of the concrete cover versus concrete core (Figure 1), except in the sections outlining the need for proper site practices like curing.

Table 1: Selected requirements for durability in the European Standard EN 206.

Type of Corrosion	Environmental Conditions	Max. w/c ratio	Min. cement content, kg/m^3	Strength Grade
Carbonation	Dry or Wet	0.65	260	C 20/25
	Wet rarely dry	0.60	280	C 25/30
	Moderate humidity	0.55	280	C 30/37
	Cyclic wet/dry	0.55	300	C 30/37
Sea water chlorides	Moderate humidity	0.50	300	C 30/37
	Wet	0.45	320	C 35/45
	Cyclic wet/dry	0.40	340	C 35/45
Aggressive chemicals	Slight	0.50	320	C 30/37
	Moderate	0.45	320*	C 30/37
	High	0.45	350*	C 35/45

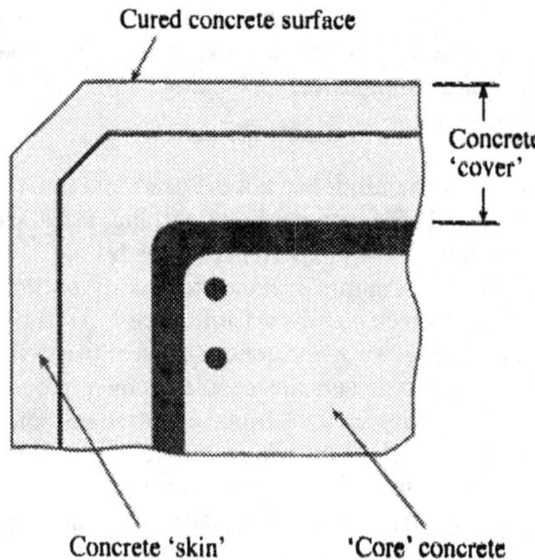

Figure 1 - Schematic presentation of the concrete cover, concrete skin and concrete core in a reinforced concrete member.

2. THE KNOWING DOING GAP

It is evident that there is a big gap between our scientific understanding of the processes leading to corrosion of concrete and reinforced concrete and the means taken in design and practice to control durability. The knowing-doing gap reflects to a large extent the philosophy of durability design which is based largely, and justly so, on means taken to produce concrete in which transport of various species is sufficiently low. The scientific concepts of producing such concretes are well known (effect of water/cement ratio, additives, degree of hydration, etc.) and therefore are not the subject of intensive investigation, whereas the implementation of this know-how into the engineering discipline is lagging. A step in the right direction has been taken in the effort in recent years to develop tools for service life quantification, which are based on the ability to model the transport characteristics. In the case of steel corrosion control, modeling which is based on simplifying assumptions using constants characterizing chloride ingress and carbonation can provide useful information and serve as an adequate guide for engineering design.

In the case of chloride penetration, simplified service life modeling can be based on the diffusion equation:

$$C(x) = C(0) \bullet \left[1 - erf \frac{d}{2\sqrt{Deff \bullet t}} \right] \qquad (1)$$

where:
C(x) – the concentration of chlorides at a depth d, after time t
C(0) – the surface concentration of the chlorides
d – depth of cover over the reinforcing bar
Deff – effective diffusion coefficient
t – time
erf – error function

In the case of carbonation the simplified semi-empirical equation can be applied:

$$d = k_c (t)^{1/n} \qquad (2)$$

where:
d – depth of carbonation after time t
k_c - a constant characteristic of the concrete and the environmental conditions
n – a parameter which is approximately constant, in the range of 2 to 3.

These equations, although simplified, can provide reasonable first order estimates of service life if the proper values of the characteristic constant (Deff in equation (1) and k_c in equation (2)) are known. Estimates for such constants and methods for their evaluation have been reviewed in several publications [e.g. 4,5].

The use of this approach can provide the tool to bridge the gap outlined above. It might be considered as an approach, which could be used within the framework of the "equivalent performance concept" which is outlined in the new European standard EU 206. The EN 206 states, "it shall be proven that the concrete has an equivalent performance especially with respect to its durability when compared with reference concrete in accordance with the requirements for the relevant exposure class".

A major technical difficulty in the application of this quantitative approach is the determination and choice of meaningful constants and boundary conditions. As already indicated, they are dependent not only on the concrete quality but also on the environmental conditions, and they may be changing over time. Procedures to take into account the influence of time have been advanced for the diffusion constants, and typical values can be found in literature [4,5]. However for these procedures to be effective there is a need for determining relevant constants reflecting the local raw materials and conditions.

3. CULTURAL CHANGE

The approach, which was briefly presented above is well known to scientists and technologists dealing with concrete. However, it is not familiar to practitioners, both in the design and construction stages. There is a need to bring across to them the significant implications of this approach and the need to develop it into working tools. The analysis and discussion in various publications, which is intended to achieve this change, is based to a large extent on qualitative statements on the limitation of strength as a quality control criteria and the need for stringent site practices to materialize the quality of the prescribed concrete.

There is a need to quantify such statements in terms, which could be readily interpreted from an engineering point of view if one wants to mobilize a change in the "culture" of addressing concrete durability by the engineering community. This can be done by calculations of first order estimates of service life based on equations 1 and 2, combined with consideration of the effect of practical variables, such as strength, curing and additives on both, strength and service life. The present paper demonstrates this approach, based on evaluation of the characteristic parameters for D (eq.1) and k_c (eq.2), using estimates obtained by accelerated tests. The parameters were evaluated by accelerated tests of chloride penetration and carbonation for a range of variables which reflect practical mix composition (w/c ratios, cement contents, fly ash additions) and curing procedures (spraying of concretes with water several times a day during the curing period and ponding for 7 and 28 days). The sensitivities of the performance from the point of views of service life and strength of the concrete cover were assessed, to quantify and demonstrate the issues outlined above. Service life was calculated from equation (1) and (2) for a concrete cover of 30mm, and the details are provided in [6,7].

4. THE VALIDITY OF STRENGTH FOR ASSESSMENT OF DURABILITY PERFORMANCE

Relations between strength and calculated service life for chloride penetration and carbonation are presented in Figures 2 for concretes with a range of cement contents of 200 to 340 kg/m^3 and curing procedures spanning from "poor" (i.e. intermittent water spray for 3 and 6 days), through "good practice" (ponding for one week) to "exceptional" (ponding for 28 days).

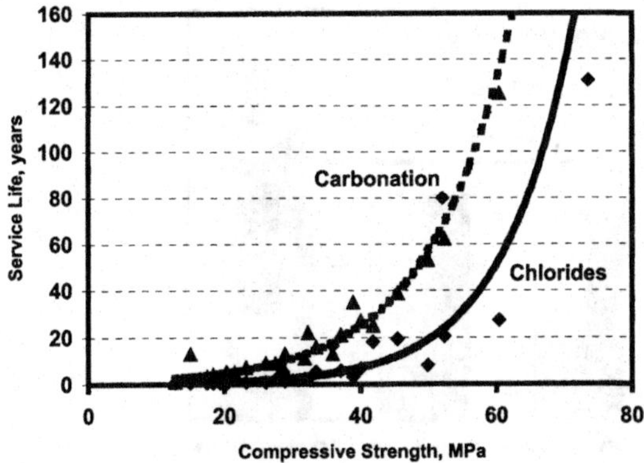

Figure 2 - Relations between compressive strength (100mm cubes) and calculated service life for chloride and carbonating conditions, assuming a cover depth of 30mm, adopted from [7,8].

The general trend of improved resistance to chloride penetration and carbonation with increased strength is clearly evident. The resulting relations clearly reinforce the perception that higher strength is linked with longer service life. It is also evident, that for chloride environment higher strength concretes are required, reflecting the prescriptions in the standards.

These are the kind of relations highlighted in textbooks and engineering guides, and form the basis of the notion that higher strength and improved durability are tightly linked. This is the concept and notion, which is bred into the civil engineering practice.

At this point, one may question how such relations are consistent with the views, which are frequently highlighted, that strength is not a good enough parameter for quantifying the concrete quality with respect to durability. The data presented in Figures 3 for carbonating environment can provide the answer to this dilemma. In this figure strength and service life values for different curing procedures (ranging from poor – intermittent spray, to exceptional – 28 days of ponding, which is the standard laboratory practice) are presented as values relative to the ones obtained by 6 days of ponding (more characteristic of good site practices). The data in this figure is for concrete required by specifications for service in harsh carbonating conditions. Similar trends were observed for chloride conditions for concrete quality required by the standards for such an environment (0.45 w/c ratio concrete).

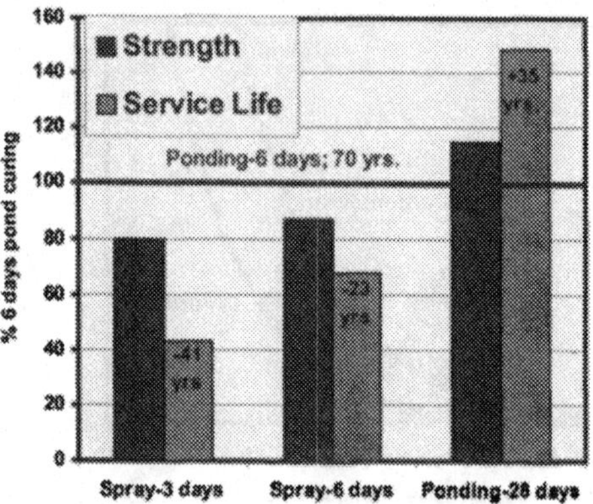

Figure 3 - The effect of curing conditions on strength and service life in carbonating conditions, relative to curing by ponding for 6 days. The concrete strength is 50 MPa (100mm cubes).

It can be seen that improved curing procedures is associated with increase in strength and service life; yet the relative effect of curing on service life is much greater than on strength. Deficient curing will result in 10 to 20% reductions in strength and 30 to 60% reduction in service life. This reduction in service life can amount to as much as 20 to 40 years, and the consequences can be estimated to be much more critical than the reduction in strength. It should be noted that the strength reduction here reflects largely the strength in the region of the concrete cover; in the actual structure the strength reduction of the core concrete would be smaller than observed here.

These observations lead to some comments reflecting the practice and laboratory testing. The enhanced service life in "exceptional curing" (the 28 days laboratory ponding curing) is indicative of the potential of a given concrete. This enhanced potential may be partially achieved only when the service conditions are wet, which is not always the case. Thus, calculation procedures in which reduction in chloride diffusion over time are applied, should be exercised with care, especially for hot and dry environmental conditions. Also, one should note that the common laboratory practices and standards to evaluate concretes which are cured under water for 28 days may provide adequate indication for strength, yet it may be misleading with respect to service life.

The data shown here can demonstrate to practitioners and specifiers, that although there is a general trend correlating between service life (i.e. durability performance) and strength, it is not sufficiently sensitive from a practical point of view: small differences in strength, which may be somehow tolerable, may be associated with much larger changes in service life, which are unacceptable, as they involve shortening of service life by more than 20 to 40 years. The data in Figure 3 can also bring across the message of the significance of curing on site, and the limitations of assessing deficient curing and its consequences by evaluating strength only.

5. STRENGTH AND DURABILITY ENHANCEMENT WITH SUPPLEMENTARY CEMENTITIOUS MATERIALS

Incorporation of supplementary cementitious materials in general, and fly ash in particular, has become a common practice in concrete technology to achieve improved durability. This technology is not sufficiently well anchored in the standards, since the prescriptive approach is not valid for it. This is reflected in the equivalent performance clause which was recently included in the European Standard EU 206. The k - efficiency factor concept introduced in the standard is also a manifestation of a performance approach. Yet, the validity of the k factor concept is being questioned, since there is no scientific backing to justify characterization of a supplementary material by a single k factor, especially when it is based on strength considerations.

This is demonstrated in Figure 4 below, which shows the efficiency factor for a particular level of replacement, calculated from the data in [6]. It can be seen that this factor is roughly constant for strength, regardless of the curing method. Yet, it varies by almost an order of magnitude when durability is considered. This difference in behavior is the result of the fact that in concretes with fly ash, the durability performance is much more sensitive to curing than the strength performance.

Data of this kind highlights the need to reconsider the validity of the efficiency concept, as well as the care that should be taken in the application of the equivalent performance concept. Tests for evaluating equivalent efficiency should consider the effect of curing; conclusions based upon standard curing of 28 days may be grossly misleading from the point of view of durability, although they may be adequate from the point of view of strength.

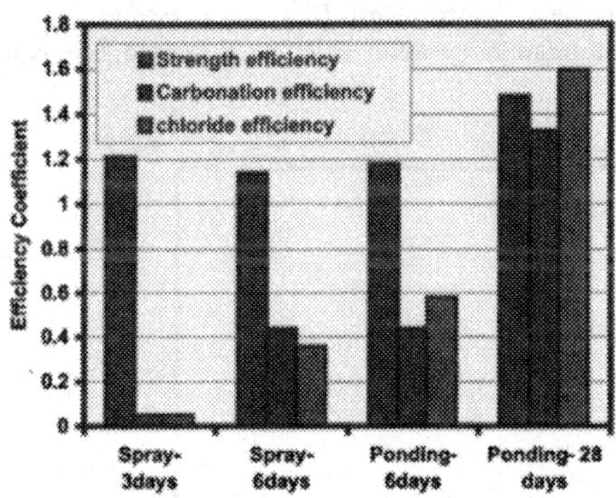

Figure 4: Efficiency factor for strength and service life in chloride and carbonation conditions, for a base concrete mix with 270 kg/m^3 cement with replacement levels of 40kg/m^3 cement by fly ash.

6. ALTERNATIVE METHODS FOR QUALITY CONTROL AND SPECIFICATIONS

The previous sections demonstrated the limitations of strength for assessing the quality of concrete, and the validity of the use of parameters which are directly related to durability performance. Service life calculations of concrete cover based on these parameters show the quantitative differences of the influence of practical variables on strength and on durability performance.

It is well recognized that in order to be able to deal with the consequences of such differences in practice, there is a need also to develop test methods, which could serve for specifying and for quality control.

At this stage, tests like the ones applied here are adequate for routine laboratory evaluation. They are sufficiently effective for providing guidelines which could be used for deign of concrete service life (if parameters required for the calculation are resolved to reflect local conditions), as well as demonstration of the need for adequate site practices, such as curing. They can highlight and quantify detrimental influences if such site measures are not taken. However, in order to have an effective quality control system, there is a need for adequate site tests, to measure characteristics other than strength, which will better reflect the durability performance. Such tests would enable testing of the actual concrete cover properties. This need is well appreciated, and a number of tests, such as surface water absorption and surface air permeability have been developed. Unfortunately, each of them has its own drawbacks, and none of them is considered as satisfactory from a practical point of view.

In order to demonstrate the validity and the potential of such a testing approach some data obtained with an air permeability test (Torrent test [8]) are presented in Figures 5 for carbonation, showing an excellent relation between calculated service life (eq. 2) and air permeability values.

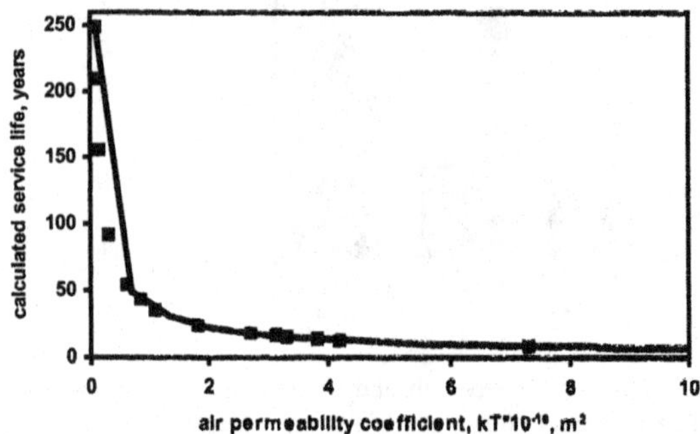

Figure 5: Relation between the calculated service life in carbonating conditions and air permeability values (adoption of data from [7]).

Similar correlations were developed for chloride environment, but they were not as good as those obtained for carbonation. First, the correlation between air permeability and chloride permeability (determined by the rapid chloride permeability test) showed some scatter. Second, the estimation of the service life from the permeability value, is dependent on the cement content of the mix, and therefore for each cement content level there is a different correlation.

Relations of the kind demonstrated here show the potential for direct evaluation of the performance of the concrete skin with respect to durability, in terms which could be readily interpreted by the engineering community. Obviously, much more work and development of instrumentation need to be done before such quantitative curves can be established with sufficient reliability. However, even at this level of know-how, some semi-quantitative parameters can be resolved.

7. CONCLUSIONS

1. From the practical point of view, the most efficient engineering tool to control durability performance, in particular for the major issues of steel corrosion and sulfate attack, is the permeability of concrete. The know-how of making impermeable concrete is well established but unfortunately it is not rooted effectively in design and field practice.
2. There is a need to re-educate the design and construction engineers to take all the necessary steps to achieve impermeability and highlight the limitations of the strength criteria when considering durability.
3. One of the ways to bring this message across is to demonstrate the differences in relative effects of curing and strength on durability performance.
4. In order to implement this approach in practice much more effort should be given to the development of tests for evaluating the performance of the concrete cover.

REFERENCES

1. A.Neville, "Consideration of Durability of Concrete Structures: Past, Present and Future", *Materials and Structures*, **34** [236] 114-118 (2001).
2. P.K.Mehta and R.W.Burrows, "Building Durable Structures in the 21st Century", *Concrete International*, **23** [3] 57-63 (2001).
3. S.P.Shah, K.Wang and W.J.Weiss, "Mixture Proportioning for Durable Concrete: Challenges and Changes", *Concrete International*, **22** [2] 73-78 (2000)
4. P.B.Bamforth, "Specification and Design of Concrete for the Protection of Reinforcement in Chloride-Contaminated Environments", Paper presented at 'UK Corrosion &Eurocorr 94', Bournemouth International Centre, 31 October – 3 November, UK, 1994.
5. A.Bentur, S.Diamnond and N.S.Berke, "Steel Corrosion in Concrete", E&FN SPON, UK, 1997.
6. A.Bentur and H.Baum, "Durability of Fly Ash and Microsilica Concretes", Research Report, National Building Research Institute, Technion – Israel Institute of Technology, Haifa, 2000.
7. I.Wasserman and A.Bentur, "Criteria for the Efficiency of Curing Concrete Structures", Research Report, National Building Research Institute, Technion – Israel Institute of Technology, Haifa, 2002.
8. R.Torrent, "A Two-Chamber Vacuum Cell for Measuring the Coefficient of Permeability to Air of the Concrete Cover on Site", *Materials and Structures*, **25** 358-365 (1992).

CONCRETE DURABILITY FROM A HIGHWAY PERSPECTIVE

Stephen W. Forster
Federal Highway Administration
McLean, Virginia, USA

ABSTRACT Although the highway infrastructure includes both pavements and bridges, for the purposes of this discussion, concrete pavement is chosen as the focal point. Durability is considered to be those aspects of the pavement behavior that impact its service life and performance during that service life. The three broad areas of performance problems for jointed concrete pavements are structural cracking, surface wear and polish and concrete deterioration. The various factors (in particular the materials factors) that cause or contribute to these problems are briefly discussed. The program of research and technology that the Federal Highway Administration (FHWA) has recently proposed is outlined as it relates to the workshop objectives. Finally, some specific activities are listed that will potentially contribute to the workshop goals.

1. INTRODUCTION

As used by transportation agencies, the term highways includes both the pavement and the bridge segments of the system used by motor vehicles. In this sense, the durability of concrete as used in highways would include its performance in both these applications. In order to simplify this discussion, only the durability of pavements, and in fact only the most commonly used concrete pavement type (jointed plain concrete pavement (JPCP)), is examined herein. Within those constraints, what is considered durability for JPCP? For a pavement engineer, it is essentially the length of time the pavement will last, or provide acceptable service; in essence, its service life. As we shall see in the following discussion, this might include a number of characteristics not typically considered as durability issues by the concrete materials engineer.

2. PAVEMENT DURABILITY

Based on the above introduction, pavement durability may be divided into three categories: the concrete's resistance to the chemical and physical aspects of the environment; the pavement's structural resistance to the loads resulting from traffic and ambient changes; and the durability of the pavement surface under the passage of vehicle tires. The characteristics of the cement and concrete used influence all three areas of durability.

Concrete's Resistance to the Environment

Concrete's resistance to the environment is influenced by the properties of the concrete typically considered when concrete durability is discussed, as in ACI Committee 201's documents. It is a material durability (rather than a structural) problem, a failure of the concrete to withstand the chemical and physical environment in which it is placed. Familiar causes of lack of this type of durability are freeze-thaw cycling, alkali-aggregate reactivity, sulfate attack or combinations thereof.

Assuring this type of durability in new construction involves a process of materials selection and proportioning in order to obtain a concrete resistant to the distresses of concern for a specific project

Resistance to Structural Load

Concrete is strong in compression, and relatively weak in tension. Therefore, any uncontrolled full-depth cracking that occurs in concrete pavements is normally due to tensile stresses exceeding the tensile strength (flexural strength) of the concrete. This situation can occur when the concrete attempts to contract (due to hydration volume decrease, or overall temperature or moisture drop) and is restrained; attempts to curl or warp (due to temperature or moisture gradients in the slab) and is restrained (due to the mass of the slab), or has inadequate or non-uniform support for applied traffic loads. As used here, this category of cracking is distinct from cracking due to lack of durability of the concrete as a material, which is included under the category of concrete deterioration, as described above.

Achieving this type of structural durability also depends on the properties of the concrete. Volume change of the concrete in response to temperature or moisture changes is influenced by the materials, and their proportions, used in the concrete. Tensile strength can certainly be increased by changes in mix design, however it cannot be adjusted enough to compensate for excessive transverse joint spacing or other design flaws.

Resistance to Surface Wear

The surface of the concrete must be able to withstand the wear and polishing actions imposed by the vehicle tires as they pass. The amount of wear and polishing that occurs will vary greatly according to the size of the vehicle, and whether it is accelerating turning or braking.

The durability of the surface greatly depends on the properties of the concrete. Properties of importance include the concrete strength, and volume percentage and hardness of the aggregate exposed in the surface zone of the pavement.

3. 2002 WORKSHOP GOAL

Based on the recent (10/30/02) guidance from J. Gebauer and J. Skalny, the remainder of this paper will consider some things that have the potential "to improve the durability of concrete world-wide", particularly in highway applications.

Current Knowledge and Advances Needed

We certainly know a great deal about concrete durability and its resistance to a range of distresses; but do we need to know more? Why do some concretes that don't meet the widely accepted air void system criteria for freeze-thaw cycling resistance perform well in the field? Do we need better tests or criteria? The rapid mortar bar test for aggregate ASR susceptibility is widely used as a means to determine concrete components and their proportions for ASR resistance. Is this the proper use of this test, or is it mainly used because it is rapid? Although DEF is not a concern in paving concrete, it can be an issue for some pre-cast bridge elements. Is there a minimum temperature we can agree on that the concrete must attain (over what time frame) in order for DEF to be a concern? Under what conditions and concrete quality is external sulfate attack a concern? Do we need better tests? Under what conditions is carbonation a problem for concrete durability? Does it affect only lower quality concrete? Is there an appropriate test?

FHWA Activities

There are certainly durability issues where we don't have all the answers. In this section, a few activities in the area of concrete durability being conducted by the FHWA will be briefly described. More information on these and other activities is available for those that are interested on our web site at www.tfhrc.gov.

In the area of resistance to freeze/thaw distress, we are continuing to conduct staff work investigating the durability of concrete with so-called "marginal" air void systems. The air void systems are being quantified via ASTM C457, then the F/T durability measured in ASTM test C-666.

A test and/or criteria are still needed to adequately evaluate the ASR potential of a job-specific concrete, including the intended proportions and aggregate gradations. The test must be relatively rapid, as well as test a concrete duplicative of the concrete to be used. FHWA staff work is currently evaluating various modifications to accelerate the concrete prism test. The FHWA labs and others are working to attain this "holy grail" of the ASR arena.

As a means of technology transfer, FHWA recently completed a contract with Michigan Technological University that resulted in a set of guidelines for evaluating materials related distress in concrete pavements (2,3,4). The guidelines provide a decision tree approach to conducting distress surveys, sampling the pavements for testing, preparing test specimens from the samples, conducting tests and reaching conclusions as to the likely cause(s) for the distress.

FHWA also recently completed a project with the National Institute of Standards and Technology (NIST) that investigated the relationship between concrete sorptivity and durability (5,6). As a result of this work, a test method was recommended which is currently going through the ASTM approval process.

Technology Transfer

A recent example of an FHWA technology transfer package is given above. Do we need better means of technology transfer? I submit that our past record of success, even with some very viable technical advances/solutions would indicate that better/new approaches to technology transfer are needed. FHWA is currently re-organizing their program of Infrastructure Research and Technology (1), and one of the key elements is the deployment process. Deployment is envisioned as including the activities of:

- Technical assistance
- Test and evaluation
- Demonstration projects
- Workshops
- Documentation to show the benefits
- Construction

to enable the transfer of research results from state-of-the-art to state-of-the-practice.

At the same time, FHWA sees the need to educate the workforce to enable them to understand, evaluate and adopt the new technology being developed. This could involve both additions to curricula at colleges and universities to prepare our new crop of materials scientists and engineers, as well as training courses for the existing workforce.

REFERENCES

1. Federal Highway Administration, Delivering Infrastructure for America's Future, 2002, publication FHWA-RD-02-091, brochure, FHWA, Washington, D.C.
2. Van Dam, T.J. et al, 2002, Guidelines for Detection, Analysis and Treatment of Materials-Related Distress in Concrete Pavements – Volume 1: Final Report, Report FHWA-RD-01-163, FHWA, Washington, D.C., 194p.
3. Van Dam, T.J. et al, 2002, Guidelines for Detection, Analysis and Treatment of Materials-Related Distress in Concrete Pavements – Volume 2: Guidelines Description and Use, Report FHWA-RD-01-164, FHWA, Washington, D.C., 236p.
4. Van Dam, T.J. et al, 2002, Guidelines for Detection, Analysis and Treatment of Materials-Related Distress in Concrete Pavements – Volume 3: Case Studies Using the Guidelines, Report FHWA-RD-01-165, FHWA, Washington, D.C., 127p.
5. Bentz, D.P. et al, 2002, Transport Properties and Durability of Concrete: Literature Review and Research Plan, Report FHWA-RD-00-073, FHWA, Washington, D.C., 63p.
6. Bentz, D.P. et al, 2002, Service Life Prediction Based on Sorptivity for Highway Concrete Exposed to Sulfate Attack and Freeze-Thaw Conditions, Report FHWA-RD-01-162, FHWA, Washington, D.C., 62p.

THE EUROPEAN APPROACH TO STANDARIZATION IN DURABILITY

Carmen Andrade
Institute of Construction Science "Eduardo Torroja", CSIC, Madrid, Spain

1. INTRODUCTION

The main characteristic of the Code System in Europe is derived from the fact that the safety of structures is considered a in several countries governmental issue. Although only few of actual structural codes are at present obligatory, the tradition of considering the importance of structural safety is still impregnating how the engineers, architects, material suppliers and contractors approach the aspects contained in the Codes. In addition, the Codes are used for teaching as well as for practice and therefore they are a key document of reference for the profession.

In the Directive of Construction Products issued by the European Commission in 1989 (1) it is considered a double possibility depending on whether the product is traditional, that is, has already or may have a standard in a reasonable short period of time, or the product is "innovative", "that is, it is not covered by a standard in the majority of European countries. In both cases the products will be marked with the CE mark for free circulation across Europe.

For innovative products or for those not having a standard, the procedure is to deliver a "Technical Approval" in which the evaluation of durability is an important aspect to be tested. The European organization for Technical Approvals, EOTA is the body responsible of this CE mark. For the so called traditional products, CEN is the organization preparing the standards.

Not all the standards, and not all their parts, will be harmonized at European level. In some cases, the CE mark applies to only some aspects of the standard. EN-206: Concrete and EN 1996 part 2: Concrete Structures (Eurocode 2) are standards and therefore voluntary specifications, although in some countries the national Codes are obligatory. Eurocodes have as its objective at present to unify criteria, although national codes are still possible to exist with National Annexes outlining the Eurocode 2 differences.

2. DURABILITY ASPECTS IN EUROCODE 2

In chapter 4, Eurocode 2 (2), named as CEN-1996 part 2, defines that: " a durable structure shall meet the requirements of serviceability, strength and stability throughout its intended working life without significant loss of utility or excessive unforeseen maintenance". Service life is usually defined as the "period of time in which the structure maintains its design requirements of : *safety, functionality and aesthetic* without unexpected costs of maintenance.

In EC-2 chapter 4 after the definition of some general principles, is included as environmental action, an exposure classification. To stand the predefined environments, Eurocode 2 only specifies the cover thickness needed to resist such environments, depending upon the type of structure, together with the maximum crack widths allowed which are specified in chapter 7 of the EC-2.

The main subchapters of chapter 4 are the following:

- definition of durability and principles
- environmental conditions: corrosion, freeze-thaw, chemical attack
- exposure classification
- requirements for durability
- methods of verification: cover thickness and allowance for deviation.

In addition of cover thickness and maximum crack width, EN-206 (3) defines the minimum cement content, the maximum w/c ratio and the need or not of air-entrained, as well in function of the predefined environmental classes. Therefore, actions are classified in types of exposure aggressiveness and resistance to each exposure aggressiveness a cover thickness of certain minimum quality is defined.

Code format and durability design

Present durability requirements (concrete mix proportioning or mechanical strength) minimum, cover thickness and maximum crack widths, fit well in a code format except that neither safety factors have been established nor specified, and that these specifications are insufficient to ensure a predefined service life, mainly when the structure is located in chloride contaminated environments.

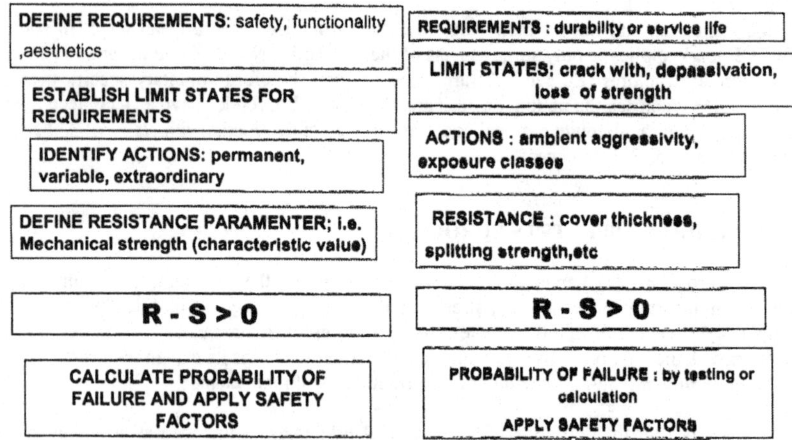

Figure 1. Format of present Codes for design based in mechanical strength (left) and the parallel format proposed for durability design (right)

Therefore, from present situation based on the mentioned deemed-to-satisfy rules, for durability to be part of the design of the structure, it has to be formulated following the format used by codes for mechanical design. That format can be summarized as follows:

- Define *requirements of safety, functionality, aesthetics*
- Establish a *limit state* for these requirements
- Identify *actions* (s): *permanent, variable, extraordinary*
- Define the parameter encountering *resistance* (R)
- Formulate the limit state function: R – S>0
- Establish the safety level of the structure
- Apply safety factors for R and S in the form: $\frac{R}{\gamma_R} \geq \gamma_S S$

Being this the rational of present Codes, durability of the structure should be treated in the same manner, and therefore a parallel methodology has to be established as is shown in figure 1.

Multilevel methodology for durability design

The main reason why strength is not enough for characterizing durability is based on the fact that it may be to a degree proportional to porosity, and therefore to permeability, it does not account for the binding of the aggressive ions (e.g., carbon dioxide, chlorides or sulphates) which suppose to act as "retarders" of the penetration process. In the case of non-diffusion controlled deleterious processes, such as alkali-aggregate or freeze-thaw, strength is not a good parameter to indicate durability either. Therefore, present durability specifications have to be complemented or changed to deal with a reliable design for durability.

A so named "multilevel approach" was proposed recently for the particular case of reinforcement corrosion (4) in order to achieve a comprehensive treatment of the problem. Basically, this multilevel approach consists of establishment of four levels of calculation or treatment of durability issues, as is shown in Table 1.

Level	Method
0	Present Codes
I	Comprehensive parameter (electrical resistivity)
II	Square root of time
III	Diffusion Coefficient

Table 1. Multilevel methodology for durability design

The level "0" is that of present Codes, based in deemed-to-satisfy rules related to concrete mix proportions and mechanical strength. The other three levels (I,II and III) proposed are based in the accounting of diffusion processes in different manners.

Level I will use electrical resistivity as the key parameter to define durability, level II will use the velocity of penetration of aggressive (either carbon dioxide or chlorides), $V_{CO2,Cl}$, and the square root of time for making the predictions, and level III will make use of he Diffusion coefficient of the aggressive. The three levels are therefore based in the same assumptions, and only changes the parameter used to account for the rate of aging and therefore, the type of test to be used in each level. In all cases the code format can be fulfilled and safety factors have to be calculated.

Level I

It is proposed to be based in the measurement of the "apparent electrical resistivity" defined as the electrical resistivity of concrete specimens water saturated at 28 days of wet curing (those used for mechanical strength), multiplied by a reaction factor r (accounting for aggressive binding)(figure 2)

$$\rho_{ap} = \rho_{28d} \cdot r \qquad [2]$$

This definition is parallel to that of the Apparent Diffusion Coefficient, D_a, which is the coefficient at a certain period of time obtained by fitting in the penetrated profile of the aggressive (usually chlorides). The advantage of the use of the Resistivity instead of that of D, is the simplicity of testing as the ρ can be measured non-destructively, either in specimens or in the real structure. Being the test non destructive, it can be easily used for quality control.

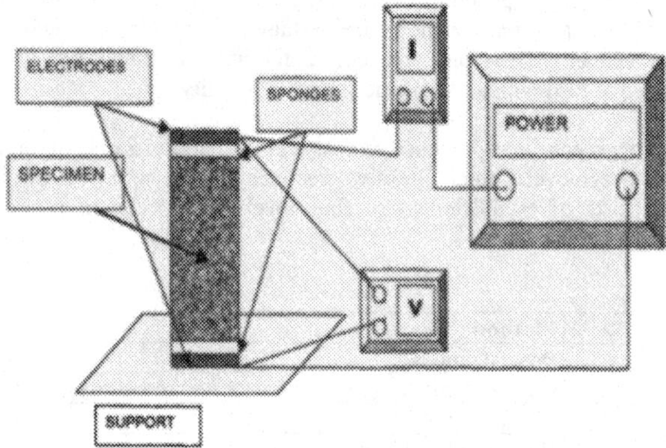

Figure 2. Arrangement for measuring the electrical resistance of concrete specimens.

Finally, it has to be mentioned that a Resistivity test is the so called "Rapid chloride permeability test" (5) although in the AASHTO or ASTM standards, this test does not takes into account the aggressive binding, which it has been however taken into account in the proposal of the "Apparent Resistivity" ρ_{ap}.

Level II

It is based in the simplified diffusion law so called "square root of time": $x= V\sqrt{t}$. In this expression the term V, is a velocity (mm/√year) and can be used to specify concrete qualities regarding durability. That is, specimens are submitted to contact with the aggressive during a certain period of time and the penetration depth of the aggressive front is measured (either with phenolphthalein for carbonation or a certain predefined chloride threshold in the case of chloride testing).

With both x and t of the test, a $V_{CO2,Cl}$ is calculated which has to be smaller (including its safety factor, γ_D) than the velocity calculated from the nominal cover and the intended service life. That is:

$$\gamma_D \cdot V_{test} = x/\sqrt{t_{test}} \quad \leq \quad V_{target} = \text{cover} / \sqrt{\text{service life}}$$

Level III

Level III is proposed for using any kind of model, generally based in the measurement and assumption of diffusion coefficients. In this level, the problem is not the type of model, as they are numerous already available in the literature (analytical, numerical, semi-empirical, etc) but: a) how they reproduce the physico-chemical processes developing in the reality and b) which is the kind of test reliably reproducing long term conditions.

There is at present a lack of calibration of existing theoretical models in three main aspects :

1) one is on the ability of these available models to reproduce the real physico-chemical reactions with the complexity that supposes that concrete is generally exposed to changing atmospheric conditions.

2) The second is the ability of models to reproduce concrete aging and the evolution of the microstructure leading to changes in the ability to resist the penetration of the aggressive

3) And the third one is that the concrete has to be tested to verify its fitness for use and therefore, the models are fed with data of tests that must be related to the models.

In consequence, level III, based in advanced models should be applied only to special structures, with very well defined environmental conditions and able to afford for large prestesting program, and the owner of the structure should be informed on the lack of full certainty on the ability of existing models to predict long term performance.

On the other hand this level is needed to accommodate research and advanced models, which have to be started to be experimented and calibrated.

3. THE EXPERIENCE OF ORESUND BRIDGE DESIGN FOR DURABILITY

It is an example of compliance with test (level I). It is known that this bridge links Sweden with Denmark (figure 3) and has been recently built. It is composed by a tunnel and a suspended bridge and carries car and train traffic. The deck of the bridge was made by prefabricated elements fabricated by Dragados, Obras y Proyectos, a Spanish company (6). The elements were prepared in a plant in the south of Spain and then transported by ship to the final location.

Differently to other initiatives of durability design, Oresund bridge approach to durability was based, not in the application of advanced durability models, but on the complying to certain limiting values of a set of tests related to several possible deleterious processes expected during the service life of the bridge.

A pre-testing program which lasted more than one year was used to select the mix proportions able to fulfil the preset limit values of the durability tests. On the other hand the mixing and curing of the concrete mixes was always predetermined in a very rigorous manner in order to control thermal cracking and shrinkage characteristics. So, the maximum allowed temperature during mixing and curing (during a minimum of 21 days) was of 60° C. A very controlled and care curing was ensured in a special building specifically built for this purpose.

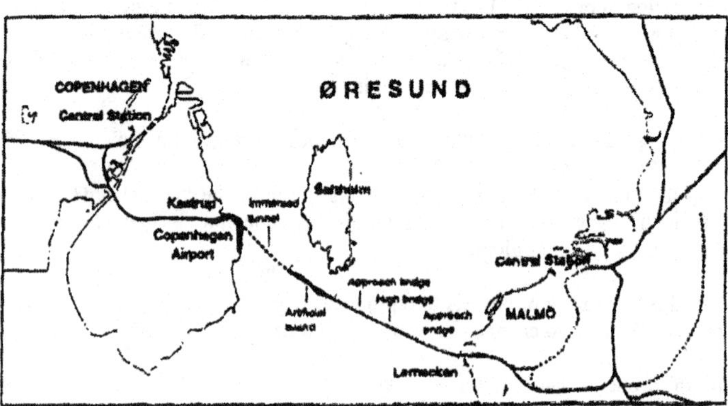

Figure 3. The Oresund Link between Copenhagen-Denmark and Malmo-Sweden

After the curing, the test program consisted in tests of:

- alkali-aggregate reaction
- chloride penetration resistance
- entrained air proportion
- scaling resistance
- shrinkage and creep

4. IMPORTANCE OF STRUCTURAL DESIGN IN DURABILITY

It is very important not to loose the perspective that the decisions related to durability may affect very much the structural behaviour and, on the opposite, the form and structural detailing may be very relevant for the durability.

Briefly, but stressing the point, it is important to realize that in general increasing durability have associated higher mechanical strengths, and therefore other stiffness, microcracking generation, etc. This means that the size and rebar detailing has to be changed from a structural point of view. These changes may mean that covers have to be reduced as a consequences of the lower concrete sections needed to support the same loads but with higher concrete strengths. In consequence, an increase in mechanical strength due to durability requirements has to be very carefully considered by the structural engineer in order not to change the structural design in such a way that the improvement in the material properties is cancelled by the necessary change in the structural design.

On the opposite, there are cases of structures that in spite their small cover thicknesses and concrete sections, they are showing very long lives in good shape. That is the case of several of the structures designed by E. Torroja (figure 4) in general membranes and shells builts during the 30's of XX Century, they present excellent durability in spite of some of them have been submitted to marine environments, were made with relatively porous concrete and have very low covers. This indicates the importance of the structural design from the view point on how the structure carries the loads or on how their equilibrium is achieved.

Figure 4. Outside (left) and inside (right) of the Algeciras Market designed by Torroja and Sanchez-Casas.

5. FINAL COMMENTS

The unsatisfactory present durability of some concrete structures located in very aggressive environments and need to design for very long service lives indicate that he requirements specified in present codes and standards are not enough.

The future evolution of these codes and standards has been proposed to be based in present communication in a "multilevel methodology", in which each level supposes a certain higher degree of sophistication in the tests procedures and the associated model for predicting service life. In any of the levels, a typical limit state function expressed as R-S ≥ O with safety factors associated, has to be complied.

From the three levels of durability design proposed, one is based in the measurement of electrical resistivity, which is a very inexpensive and a non-destructive test type very adequate for laboratory and on-site quality control. The other two are based in more specific test types associated to models accounting for service life of reinforcements.

The main steps that a code format requires for durability design are:

- Define requirements (durability, service life)
- Identify actions (S) ambient aggressivity exposure classes
- Define Resistance (R) parameters: resistivity, penetration velocity, diffusivity
- Define limit states: cover thickness, crack with, depassivation
- Apply limit state function: R-S ≥ 0
- Calculate probability of failure and apply safety factors

Finally, remarks are made on the importance of having a comprehensive durability design, not only from a material point of view but also from a structural perspective. Durability design has to be integrated as an overall conception into the engineering structural design.

References

(1) EU-Directive of Construction Products.
(2) CEN-EN-1992-Part 1. Eurocode 2 – Design of concrete structures. Part 1: General rules and rules for buildings.
(3) CEN-EN-206 – Concrete – Part 1: Specifications, performance, fabrication and conformity.
(4) C. Andrade, C.Alonso, A. Arteaga, P. Tanner, "Methodology based on the electrical resistivity for the calculation of reinforcement service life" 5th Canmet/ACI Int. Conference on Durability of concrete – Supplementary papers volume. Barcelona, Spain, June (2000) 899-915.
(5) D. Whiting – Rapid determination of the chloride permeability of concrete – Federal Highway Administration – Report FHWA/RD-81/119 (1981).
(6) G. Serrano, J. Obregón, J. Rodríguez, P. Trigo, F. Hue y L. Peset, "Fabricación de tableros para el puente de Oresund entre Dinamarca y Suecia (Fabrication of decks for Oresund bridge between Denmark and Sweden) – 1er Congreso de la Asociación Científico-Técnica del Hormigón, ACHE (1st Congress of ACHE) – Sevilla – España – (1999).

II. Conclusions of the 4th Anna Maria Workshop (2003)

ANNA MARIA WORKSHOP 2003

TESTING & STANDARDS FOR CONCRETE DURABILITY

WORKING GROUP 1

MATHEMATICAL MODELING AND STANDARDS FOR PREDICTION OF CONCRETE SERVICE LIFE

Group Members:

Geoffrey Frohnsdorff (chair)	NIST (retired), USA
Jacques Marchand (co-chair)	University of Laval, Canada
Andrew Boyd	University of Florida, USA
Nick Buenfeld	Imperial College, UK
Sidney Diamond	Purdue University, USA
Carolyn Hansson	University of Waterloo, Canada
David Myers	Grace Construction Products, USA
Ken Snyder	NIST, USA
Larry Sutter	Michigan Technical University, USA
Peter Taylor	PCA, USA

1. INTRODUCTION

The objective assigned to the Working Group was "to summarize the state-of-the-art of modeling as it relates to standards needed for prediction of the service life of concrete and concrete structures, both plain and reinforced. Also, to propose methods for implementation" [of service life prediction models]. The objective is a reflection of the fact that demands for enhanced technical performance, safety, economics, and environmental protection, cause a pressing need to be able to determine at the design stage, with an acceptable degree of confidence, the projected service life of any critically-important concrete structure such as a major bridge or tunnel, or a nuclear waste containment facility. The development and standardization of this ability will also be useful in estimating the life of any new or existing concrete structure. Progress in this direction, particularly as it relates to new structures, is the subject of this report.

At present, estimates of service life ("durability") of concrete tend to be made on the basis of the designer's judgment taking account of experience with similar structures in similar environments, and the present incomplete understanding of the physical and chemical processes that degrade concrete in service in the field. However, for several noteworthy infrastructure projects, empirical or semi-empirical mathematical models expressing information about relationships between concrete composition and service life have already been used.

In the opinion of the Working Group, the development of mathematical models that are both practical and scientifically-correct is needed to enable more reliable predictions of the service lives of concrete structures. Such models will be of significant benefit to society since, as just one example, they will aid the design of important infrastructure projects by enhancing the probability that the intended life, the design life, will be achieved.

The references listed at the end of this report are to literature that helps define the state-of-the-art of mathematical modeling of service life prediction, but without any claim to completeness.

2. TYPES OF MODELS

For the purpose of clarity, it may be helpful to point out the three main categories of models for predicting the service life of concrete -- empirical, semi-empirical, and mechanistic. An empirical model is one which makes predictions based on previously observed relationships among service life, concrete composition, and the exposure conditions of structures, without invoking understanding of the scientific reasons for the relationships; this category includes neural network models. On the other hand, a mechanistic (or physico-chemical) model provides predictions of service life based on a mathematical formulation of relevant aspects of the internal structure of the concrete, and of the specific changes in it induced by the various physical and chemical mechanisms of deterioration that may be encountered in service. As long as the internal structure will not change significantly in the period of interest, the relevant aspects of the internal structure may be represented in a simplified form such as effective porosity. A semi-empirical model combines features of mechanistic and empirical models. In principle, empirical and semi-empirical models are restricted to providing predictions for concrete structures and environmental conditions (i.e. boundary conditions) in ranges for which some service life data are already available. When comprehensive mechanistic models are developed, they should be essentially free from such restrictions -- they should be able to predict the service life for any concrete structure for any boundary conditions consistent with the inherent mechanistic assumptions within the model. However, mechanistic models, like most other models, result from an evolutionary process in which they become increasingly comprehensive and rigorous as new scientific knowledge is incorporated.

An ideal service life prediction model would be probabilistic, with the predicted service lives being expressed in the form of probabilistic distribution functions. In principle, the ideal model should be able to estimate the magnitudes of uncertainties that may be caused by inhomogeneities [1], including cracks and other defects, whether due to poor workmanship, inadequate curing, extreme environmental exposure conditions, or some other cause.

3. MATHEMATICAL MODELS AND ADVANCES IN ENGINEERING

The influence of mathematical modeling on the advance of technology is apparent in many fields. To give a few examples, mathematical modeling has contributed substantially to progress in design of aircraft, spacecraft, and ground transportation systems; in nuclear and chemical engineering; and in the design of long-lived concrete barriers for the containment of nuclear wastes. Its important role in weather forecasting is also worthy of note. Modeling is invaluable in predicting the performance of complex systems with many interacting parts or processes. Concrete is such a system. While there is much to be done, the rate of development of mechanistic mathematical models capable of modeling concrete microstructure and predicting concrete behavior is accelerating [2,3,4,5]. However, complexities of the internal structure of concrete and of the degradation processes affecting concrete in service make service life prediction a demanding application for such models [6,7].

4. STATE-OF-THE-ART OF MODELING OF CONCRETE SERVICE LIFE

As can be seen from the references in Section 8, good progress in modeling the microstructure and performance of concrete is being made at research institutions in the U.S. [8,9] and other countries [10,11,12] but, as yet, there are no standardized models for prediction of service life of concrete. Nevertheless, a number of existing models are advancing concrete technology by providing insights into quantitative relationships among factors affecting service life that could not be obtained so effectively in other ways [13,14,15,16]. Models have been used to aid the design of structures that are required to achieve a specified long life such as 100 years for a bridge or tunnel, or 500 years for a nuclear waste containment structure.

To show the complexity of the concrete service life prediction problem, Table 1 has been drawn up. It indicates, for each main type of degradation, types of concrete structure in which it is most frequently encountered. For each type of structure listed, "**O**" indicates a degradation process that is encountered frequently, and "**o**" indicates a process that is sometimes encountered. The footnote to the table, Note 1, points out that there are similarities among the degradation processes, in that most require one or more of transport of water or ions within, or in or out of the concrete, causing microstructural changes which cause deleterious changes in the mechanical properties of the concrete. The similarities draw attention to the importance of being able to model transport properties and processes [17], but this is not enough in view of the differences among the degradation processes, in the ions involved, and in the sequences of events that cause degradation.

In spite of the encouraging progress being made, it must be noted that, at present, there are several hindrances to the development and exploitation of concrete service life models. The hindrances include:

a) distrust of models because their complexity may make them difficult for a non-mathematician to comprehend, and because unrealistic claims have been made for the usefulness of some models;

b) inability of many current-generation computers to handle such large and complex models;

c) distribution of model development activities among several centers, with levels of financial support that are not consistent with the potential large benefits;

d) lack of established methods for characterizing some aspects of concrete materials and concrete in the required detail;

e) gaps in understanding of the internal structure of concrete;

f) lack of the large and comprehensive databases that would facilitate the testing of models, particularly mechanistic models;

g) lack of reference concrete structures that could provide well-documented data over many decades for inclusion in the databases; and

h) the frequently narrow scopes of, and weak interactions among, concrete standards committees that tend to inhibit the integration of knowledge that models bring about.

5. SOME CENTERS OF CONCRETE SERVICE LIFE MODELING ACTIVITY

Among organizations that are carrying out, or supporting, modeling activities related to service life of concrete are, Princeton University, NIST, and the FHWA in the U.S.; Laval University in Canada; the University of Tokyo in Japan; Imperial College in the U.K.; a number of laboratories in the European Union (EU) that have participated in, or are participating in, EU programs such as the DuraCrete program and the CONlife program; and, also in Europe, Working Group 11 of MACSI (Mathematics, Computing and Simulation for Industry). There are also some related modeling activities concerned with economics and with environmental impact that can be related to service life, whether of concrete or other materials. Web addresses for some of these organizations are listed in Table 2.

6. POSSIBLE STEPS IN THE IMPLEMENTATION OF SERVICE LIFE MODELS

There are no current standards for models for predicting the service life of concrete. However, because models and model-based calculations are making significant contributions to the advance of concrete technology, standards are needed to provide confidence in their soundness and their ability to perform as claimed. Establishment of the standards will probably require a change in the mind-set of those accustomed to developing predominantly empirically-derived, prescriptive standards for concrete and concrete materials, rather than standards for material-science-based predictions of service life and related aspects of performance. While establishment of new standards is likely to be an essential step in the movement towards use of predictive models, it is unlikely, by itself, to bring about their early acceptance.

Table 1. Types of Degradation Encountered in Plain and Steel-Reinforced Concrete in Some Structural Applications ("O" indicates frequent occurrence, "o" less frequent.)

Type of Degradation / Application of Concrete	Chloride-Induced Corrosion	Carbon Dioxide Reaction	Sulfate Attack	Alkali-Aggregate Reaction	Freezing and Thawing	Surface Scaling	Leaching of Soluble Phases	Magnesium Salt Attack
Marine structures	O		o	o	o	o	o	o
Highway pavements			o	O	O	o	o	o
Bridge decks	O	o		o	o	o		o
Foundations			O	o	o			o
Above-grade buildings		O		o				
Dams				O	o		o	
Tanks and pipes	o	o	o	o	?		O	o
Tunnels	O	o	O	o		o	o	o

Note 1. Each type of degradation involves most, or all, of the following: i) transport of water and ions within, and into or out of, the concrete, ii) a reaction, usually involving a volume change, and iii) deleterious alteration of the internal structure of the concrete. The details can differ widely. The processes may involve a chemical, thermal, mechanical, or electrochemical response, or some combination of these.

Standards for concrete and concrete materials are normally either prescriptive or performance-based, though some are hybrid -- part prescription, part performance. Predictive standards [18] may be thought of as a class of performance standards and, at present, use of performance standards is seldom favored by specifiers. Nevertheless, because of their potential advantages, it has been suggested [19] that mathematical models for service life prediction will come to be accepted over time through increasingly frequent occurrences of sequences of events roughly as follows:

a) An owner sets the service life required for a new concrete structure, and describes the environment to which the structure will be exposed.

b) The owner and the designer agree on a relationship, probably supported by mathematical models, between the results of "index tests" and the likely performance of the structure (for instance, a concrete pavement of the proposed design containing 6% entrained air is likely to survive 50 years of salting and severe winters) -- see Note 2.

c) The specification sets the tests and limits for the project so as to provide an acceptable probability that the required service life will be achieved.

d) The contractor and the ready-mixed concrete supplier agree, subject to prequalification, on a concrete system, including materials, proportions, and workmanship details. The system is pre-qualified using the tests selected in Step 3 on mock-ups, or on previously-constructed systems. (Some of the tests may take extended periods of time. For instance, the ASTM C1293 test for alkali-silica reactivity takes two years.)

e) Tests are selected for QA/QC (quality assurance/quality control) because they are rapid and are known to correlate with those used for pre-qualification; for example, ASTM C1260 for alkali-silica reaction takes only two weeks.

f) QA/QC activities employ the selected tests to prove that the concrete system in place is equivalent to that which was pre-qualified.

g) QA/QC testing is required at each change of ownership in the supply chain of the concrete materials and the concrete (i.e., when materials are delivered to the batch plant, when concrete leaves the ready-mixed concrete truck, and after curing is complete).

h) A type of warranty system may be required for "gray" areas until a sufficiently reliable set of index tests has been developed and proven.

i) As confidence grows in individual mathematical models, including models for service life prediction, the models may be substituted for some testing at the pre-qualification stage.

j) Data obtained will, increasingly, be collected in databases which, eventually, will provide support for standardization of models.

Note 2. i) This approach would require that appropriate test methods be available and limits be selected for tests for relevant degradation mechanisms, ii) the scatter in test results would have to be taken into account when selecting the limits, iii) sufficient data, or appropriate models, should be available to assure the owner of the adequacy of the correlation between tests and service life, iv) for cost-effectiveness, some tests might be prescriptive: for instance, a limit might be imposed on chloride content instead of requiring corrosion rate measurements to be made, v) other tests might be performance-based: for instance, measurement of alkali expansion of the concrete system might be preferable to imposing prescriptive chemical limits on the system.

Processes such as that envisioned will require guides to be available to assist specifiers in selecting test methods and limits for the expected service environments. In the initial stages, owners will have to accept some risk and trust that the approach is valid based on the data available at the time. (This is little different from the risks a building owner accepts at present.)

As more structures are built and historical data are accumulated, confidence in model-supported design processes will grow, as will their use.

Table 2. Web Sites for Information on Concrete Service Life and Related Models

Electronic Monograph, VCCTL, COST, CIKS, Visible Cement (NIST) http://ciks.cbt.nist.gov
Computation of the Elastic Properties of Concrete (NIST) http://math.nist.gov/mcsd/savg/parallel/epc/
BridgeLCC, Life-Cycle Cost Analysis for Bridges (NIST) http://bfrl.nist.gov/bridgelcc/welcome.html
BEES, Building for Environmental and Economic Sustainability (NIST) http://www.bfrl.nist.gov/oae/bees.html
Duracrete (an EU program) http://www.duranetwork.com/
CONlife (an EU program) http://www.cordis.lu/growth
MACSI Working Group 11, Concrete as Multiphase Porous Material http://www.win.tue.nl/macsi-net/WG/WG11.html

Recognition by standards bodies that model-based calculations will make a significant contribution to the future development of standards for performance of concrete is becoming apparent. One encouraging development is that test method subcommittees in ASTM Committee C01, Cement, have accepted the idea of attempting to develop a virtual test method (i.e., an ability to make model-based predictions) to complement each physical test method. And ACI and ISO have both established committees concerned with concrete service life prediction. In ISO, Technical Committee 71 on Concrete, Reinforced Concrete, and Prestressed Concrete, of which ACI is the secretariat, has recently established Subcommittee 7 on Service Life Planning of Concrete Structures.

To assist ISO TC71/SC7, ***fib*** (the Federation Internationale de Beton) is drafting a document on service life planning of concrete structures that it proposes to pass over to SC7 as a working document for processing as an international standard. Examples of modeling and related activities under way in ACI include Committee 365, Service Life Prediction, and those in Committee 235, Knowledge-Based Systems and Mathematical Modeling of Materials. The committee stimulated publication of a report on the benefits of computerizing knowledge of concrete [20], and a subcommittee of ACI 235 has prepared a "data dictionary" [21] to aid the development of the large databases that will be needed to provide input data to concrete modeling activities, and the ACI Board formed an ad hoc task group to look at the implications for ACI of Virtual Concrete Technology. As a general comment, it may be noted that terminology relating to mathematical modeling of the properties, performance, and service life of concrete is not well-developed, and it would be helpful to have a standard terminology to facilitate communication between model developers and model users in the concrete community.

The concrete community might obtain some guidance on development of service life prediction standards from principles incorporated in the variety of generic standards that have been issued [22] by ISO, ASTM, AIJ, BSI, and CSA. Ultimately, it is to be hoped that the essential structures of all service life prediction standards, whether or not they are material-specific, will be harmonized.

7. CONCLUSIONS AND RECOMMENDATIONS

a) Empirical and semi-empirical models have been used to aid prediction of the service lives of critically-important concrete structures for some years. Advances in information technology and in scientific understanding of concrete have made possible the development of more sophisticated models that is now taking place in many laboratories.

b) Although, at present, there are no standards for service life prediction of concrete structures, such standards are now being drafted by an ***fib*** commission.

c) Testing of the realism and accuracy of the results of service life models is a difficult task. It would be helpful to accumulate reference data sets covering case histories of well-studied structures against which to test model predictions.

d) In view of the complexity of service life prediction models, skill in describing and explaining the operations carried out by the computer code, together with the underlying assumptions, is needed by individuals and agencies that develop such models.

e) In using service life prediction models and reporting the results, failure criteria should be defined, information should be provided on how boundary conditions were determined and entered into the model, and attention should be drawn to limitations on application of the results.

f) In some cases, computer codes for concrete service life prediction models may become parts of computer-integrated knowledge systems that are made freely

g) available through the Web. In others, the codes may be considered to be proprietary information and not made freely available. Because the complexity of the codes would seem to require electronic, rather than, paper-based publication, concrete societies may need to adjust their publication policies to facilitate electronic publication.

REFERENCES

1. Garboczi, E.J., and Bentz, D.P., Effect of Statistical Fluctuation, Finite Size Error, and Digital Resolution on the Phase Percolation and Transport Properties of the NIST Cement Hydration Model, *Cement and Concrete Research*, Vol. 31, No. 10, pp.1501-1514, October 2001.

2. Bentz, D.P., Garboczi, E.J., and Martys, N.S., Application of Digital-Image-Based Models to Microstructure, Transport Properties, and Degradation of Cement-Based Materials, in *Modelling of Microstructure and Its Potential for Studying Transport Properties and Durability*, Jennings, H., Editor, Kluwer Academic Publishers, pp.167-185, 1996.

3. Bentz, D.P., Three-Dimensional Computer Simulation of Portland Cement Hydration and Microstructure Development, *Journal of the American Ceramic Society*, Vol. 80, No. 1, pp. 3-21, 1997.

4. Garboczi, E.J., and Bentz, D.P., Percolation Aspects of Cement Paste and Concrete: Properties and Durability, *High-Performance Concrete: Research to Practice.* American Concrete Institute Spring Convention. Proceedings. ACI Special Publication 189 March 14-19, 1999, Chicago, IL, pp.147-164, 1999.

5. Bentz, D.P., and Garboczi, E.J., Multi-Scale Microstructural Modeling to Predict Chloride Ion Diffusivity for High Performance Concrete, *Materials Science of Concrete*, Special Volume: Ion and Mass Transport in Cement-Based Materials. Proceedings American Ceramic Society October 4-5, 1999, Toronto, Canada, pp. 253-267, 2001.

6. Snyder, K.A., and Clifton, J.R., 4SIGHT Manual: A Computer Program for Modelling Degradation of Underground Low Level Waste Concrete Vaults, *NISTIR 5612,* 73 pp., June 1995.

7. Snyder, K.A., Validation and Modification of the 4SIGHT Computer Program, *NISTIR 6747*, 85 pp., May 2001.

8. Bentz, D.P., and Forney, G.P., User's Guide to the NIST Virtual Cement and Concrete Testing Laboratory Version 1.0, *NISTIR 6583*, 54 pp., November 2000.

9. Haecker, C.J., Bentz, D.P., Fen, X.P., and Stutzman, P.E., Prediction of Cement Physical Properties by Virtual Testing, *Cement International*, Vol. 1, No. 3, pp. 86-92, 2003.

10. Sagues, A., Model for a Quantitative Corrosion Damage Function for a Reinforced Concrete Marine Substructure, in G. Frohnsdorff (ed.), *Modelling Service Life and Life-Cycle Cost of Steel-Reinforced Concrete,* Report from the NIST/ACI/ASTM Workshop Held in Gaithersburg, MD on November 9-10, 1998, NISTIR 6327, 55 pp., May 1999.

11. Nilsson, L.O., Present Limitations in Scientifically-Based Prediction Models for Chloride Ingress Into Submerged Concrete, in G. Frohnsdorff (ed.), *Modelling Service Life and Life-Cycle Cost of Steel-Reinforced Concrete,* Report from the NIST/ACI/ASTM Workshop Held in Gaithersburg, MD on November 9-10, 1998, NISTIR 6327, 55 pp., May 1999.

12. Poulsen, E., Chloride Exposed RC-Structures: Chloride Ingress and Lifetime Prediction by the Hetek Model in G. Frohnsdorff (ed.), *Modelling Service Life and Life-Cycle Cost of*

Steel-Reinforced Concrete, Report from the NIST/ACI/ASTM Workshop Held in Gaithersburg, MD on November 9-10, 1998, NISTIR 6327, 55 pp., May 1999.

13. Bentz, D.P. and Conway, J.T., Computer Modeling of the Replacement of "Coarse" Cement Particles by Inert Fillers in Low W/C Ratio Concretes: Hydration and Strength, *Cement and Concrete Research,* Vol. 31, No. 3, pp. 503-506, 2001.

14. Remond, S., Bentz, D.P., and Pimienta, P., Effects of the Incorporation of Municipal Solid Waste Incineration Fly Ash in Cement Pastes and Mortars. Part 2: Modeling, *Cement and Concrete Research,* Vol. 32, No. 4, pp. 565-576, April 2002.

15. Bentz, D.P., Influence of Silica Fume on Diffusivity in Cement-Based Materials. II. Multi-Scale Modeling of Concrete Diffusivity, *Cement and Concrete Research,* Vol. 30, No. 7, pp. 1121-1129, July 2000.

16. Garboczi, E.J., Three-Dimensional Mathematical Analysis of Particle Shape Using X-Ray Tomography and Spherical Harmonics: Application to Aggregates Used in Concrete, *Cement and Concrete Research,* Vol. 32, No. 10, pp.1621-1638, October 2002.

17. Bentz, D.P., and Garboczi, E.J., Computer Modelling of Interfacial Transition Zone: Microstructure and Properties, *Engineering and Transport Properties of the Interfacial Transition Zone in Cementitious Composites. RILEM Report No. 20. Proceedings.* Part 5, Chapter 20, 1999, Cedex, France, RILEM Publications s.a.r.l., Cedex, France, Alexander, M.G., Arliguie, G., Ballivy, G., Bentur, A., (eds.), pp. 349-385, 1999.

18. Hill, E., and Frohnsdorff, G., Portland Cement Specifications: Performance, Prescription, and Prediction, *Cement, Concrete, and Aggregates,* Winter 1993, American Society for Testing and Materials, Philadelphia, pp. 109-118, 1993.

19. Taylor, P., private communication.

20. Frohnsdorff, G., and Kaetzel, L.J., Computerizing Concrete Technology Knowledge, *Concrete International,* pp. 74-76, December 1999.

21. Oland, C.B., and Ferraris, C.F., Concrete Materials Database, *Concrete International,* Vol. 22, No. 12, pp. 28-33, December 2000.

22. Frohnsdorff, G., and Martin, J.W., Towards Prediction of Building Service Life: The Standards Imperative, *Proceedings, 7th International Conference on Durability of Building Materials and Components,* C. Sjostrom (ed.), Chapter 147, Vol. 2, Testing, Design and Standardization,. May 19-23, 1996, Stockholm, Sweden, E. & F.N. Spon, New York, pp. 1417-1428, 1996.

ANNA MARIA WORKSHOP 2003

TESTING & STANDARDS FOR CONCRETE DURABILITY

WORKING GROUP 2

TEST METHODS AND SPECIFICATIONS

Group Members:

R. D. Hooton (chair)	University of Toronto
J. C. Roumain (co-chair)	HOLCIM
J. Lukasik	Lafarge Corporation
H. Farzam	CEMEX
N. Hearn	University of Windsor
E. Attiogbe	MBT
N. Cumming	Levelton Engineering
D. Reaves	Troxler Electronic Laboratories
A. Gee	Lehigh Cement
C. Ishee	Florida Department of Transportation
L. Suz-Chung Ko	Holcim Group Support

1. INTRODUCTION

Test methods related to measurement of various durability properties exist in various standards (eg., ASTM, AASHTO, Corps of Engineers (CRD), individual DOT's) in North America and abroad. Limits based on some of these test methods are specified in ACI, BOCA, CSA and individual DOT specifications, amongst many others. It was reported by Roumain (2002) that in the US alone there are over 2000 specifications for concrete. Each of these specifications employs different test methods and different test limits.

Another issue is that tests do not exist for all of the relevant durability or performance concerns. As well, existing tests are not always rapid, accurate, repeatable nor necessarily do they adequately represent any or all of the exposure conditions in-situ.

In some cases, such as for sulfate resistance, tests only exist which attempt to evaluate the chemical resistance of the cementing system (eg. ASTM C452, C1012) but not the concrete quality (eg. w/cm limits). Typically, concrete quality limits are contained in specifications such as ACI 318. These could be considered presumptive, but are a necessary part of current specifications.

The lack of adequate performance-related test methods for concrete is one of the main factors which inhibits the move from prescriptive to performance specifications.

Most deterioration process have two-stages. Initially, aggressive fluids (water, solutions, gases) need to penetrate or be transported through the capillary pore structure of concrete to reaction sites (eg. chloride to reinforcement, sulfates to aluminates) prior to the actual chemical or physical deterioration reaction.

An accepted test to measure transport rates, or a related index test is fundamental to development of any performance-based durability specifications.

2. TEST METHODS

Test methods were identified and grouped based on the durability/performance issue addressed. ASTM, AASHTO, CRD, USBR, and CSA test methods were identified, since they were the most familiar to the working group members. Where possible, European and other national test methods have also been included in the summaries.

The methods were tabulated by subject. The tests were broken down as follows:

1. Type and Standard and number
2. The type of specimen tested
3. The property measured
4. Strengths of the method
5. Weaknesses of the method
6. Improvements needed

Laboratory-based methods were identified as well as in-situ field tests where they are known to exist. One problem is the common lack of methodology for checking the validity of test methods and specification limits for assurance of durability.

3. PENETRATION RESISTANCE

The resistance of concrete to penetration of aggressive fluids is fundamental to its durability to most forms of degradation. In most specifications, there are no limits placed on permeability, diffusion, absorption or other indirect indicators of penetration resistance (eg. The so-called Rapid Chloride Permeability Test (RCPT), AASHTO T277/ASTM C1202). Typically, upper limits on W/CM and/or minimum strengths are specified in ACI 318 and CSA A23.1 for concrete in specific aggressive exposures. This leads to cases where, for example chloride exposures, all 0.40 W/CM, 35MPa concretes would satisfy the specification, even though a Portland cement concrete will have much lower resistance to chloride diffusion than one containing SCM's. Therefore, to provide a better measure of equivalent performance, it would be preferable to specify a measure of penetration resistance.

Until recently there have been very few standardized test methods for measuring such properties. Also, many methods are not rapid, and in the case of rapid tests, there is not always widespread acceptance due to perceived or real limitations. See Table 1.

The AASHTO T259 chloride ponding test requires almost 6 months to complete and there is no clear way provided to interpret the results (McGrath and Hooton, 1999). Transport mechanisms in this test also include undefined components of absorption, diffusion, and wick action. This test has been recently modified as ASTM C1443.

ASTM C1556 is a chloride bulk diffusion method, adapted from Nordtest NT Build 443. While it appears to be a useful method for prequalification of concrete mixtures and provides input data for service-life-prediction models such as LIFE365, it takes about 3 months to complete.

The Corps of Engineers has two tests for water permeability, CRD 48 and CRD 163. The former is only sensitive to low cement content concretes, typical of mass concrete in hydraulic structures. The latter is able to measure flow through better quality concretes and modifications to increase sensitivity have been proposed (El-Dieb and Hooton, 1995), but not incorporated in the standard. With both methods, obtaining data is sometimes problematic and slow.

The so-called Rapid Chloride Permeability Test (RCPT), AASHTO T277/ASTM C1202, originally developed by D. Whiting as a rapid alternative to AASHTO T259, has been used since 1983 and has become the most widely accepted "permeability" test method in North America as well in many parts of the world. It has also been widely criticised. It is really a somewhat awkward conductivity (inverse of resistivity) test. Concretes with high conductivity exhibit heating due to the 60V DC potential applied over the 6-hour test period. While this is not of concern for concrete with coulomb values less than 1500, the heating increases conductivity and exaggerates the coulomb values obtained for poorer quality concretes. This is the reason that it has been suggested to take the 30 minute reading multiplied by 12 to obtain a 6 hour value without the effects of heating (McGrath and Hooton, 1999; Julio-Betancourt and Hooton, 2004). Also ASTM C09.66 is currently balloting a modified version of C1202, where the 1-minute conductivity is measured and used as the figure of merit. The other problem with this test is that admixtures which increase the ionic conductivity of the pore fluids in concrete will result in unfairly high coulomb values, which do not reflect its chloride diffusion resistance. The most notable example of this is with calcium nitrite corrosion inhibitors.

While C1202 is not used in the ACI 318 specification, a limit of 1500 coulombs has been specified in the Canadian CSA S413 Parking Structures Standard since 1994. Currently, CSA A23.1(the 2005 draft) is likely to adopt a 1500 coulomb limit for reinforced concrete exposed to de-icer salts, and 1000 coulombs for HPC.

AASHTO TP64 was adopted in 2003 as a provisionary standard. It is a rapid chloride migration test based on Nordtest NT Build 492, developed by Tang and Nillson, 1991 and adapted by Stanish, Hooton, and Thomas, 2000. The Nordtest method includes calculation of a non-steady state chloride diffusion coefficient, but the AASHTO version only measures chloride penetration of chlorides as mm/Volt-hour. This is due to concerns that the theory behind the diffusion calculation are not valid (Stanish, Hooton, and Thomas, 2000). This test ranks concretes the same as with ASTM C1202, but has the advantage of of being influenced by strongly ionic admixtures, such as calcium nitrite.

Some researchers and various agencies have used resistivity tests, such as 4-point Wenner probes for evaluating both lab and field concrete. To date, none of these tests have been standardized.

ASTM subcommittee C09.66 is at the point of standardising a rate of absorption test method, which will likely be complete in 2004. This test is related to the British BS1881 ISAT test, developed by Levitt, but the procedures are better defined.

Table 1
Penetration Resistance Tests

Test Method		Lab-Based						Wenner Resistivity
		CRD 163	TP64/NT492	C 1556	C 1202	C 1543	T259	
Test Material	Aggregate							
	Paste							
	Mortar							
	PC Concrete	x	x	x	x	x	x	X
	SCM Concrete	x	x	x	x	x	x	X
Properties Measured		Water Permeability	Chloride penetration	Bulk Diffusion	Conductivity	Chloride Profile	Chloride Penetration	Surface Resistivity
Strength		Sensitive to low-permeability concretes	Rapid and chloride front not affected by pore fluid conductivity	Measures diffusion	Rapid			Rapid Can be used in Field
Weakness		Slow and can not always measure flow	Basis for diffusion value in NT Build 492 questioned	3-month test	Affected by conductive admixtures	6-month test	6-month test	Not yet standardised Affected by surface carbonation
Improvements Needed		Improve flow measurements						

4. ALKALI-SILICA REACTION (ASR)

Table 2 lists the test methods identified. In ASTM, numerous test methods exist for either, a) evaluation of aggregate reactivity, or b) evaluation of mitigative measures. Some methods can be used for both purposes. The CSA has developed a guide and flow chart for selecting tests for evaluating an aggregate's reactivity, this leading to selection of mitigative measures to minimize the risk of deleterious expansion and cracking. In the CSA Guide, only petrographic analysis (ASTM C295), the rapid mortar bar test (CSA A23.2-25A, ASTM C1260); the concretes prism test (CSA A23.2-14A, ASTM C1293) was used, but a modification of ASTM C1260 (called CSA A23.2-28A) is used to evaluate cement-pozzolan/slag efficiency. These tests are organized in the guide/flow chart outlined in CSA A23.2-27A (Thomas, Hooton and Rogers, 1997).

In ASTM, the Appendix to the Aggregate specification, C33, simply lists various test method limits, without providing any guidance as to selection of appropriate test methods. The list in C33 includes C227 mortar bars, C287 quick chemical tests, C1260 accelerated mortar bars, and C1293 concrete prisms. Various other ASTM specifications use ASTM C441 (or C227 with Pyrex glass synthetic aggregate) to determine efficacy of the pozzolan or slag (C618, C1240, C989) in controlling ASR. In C150, there is only the optional prescriptive low-alkali Portland cement limit of 0.60% equivalent sodium content.

The RILEM ASR committee is focussing on test methods similar to the CSA standard. ASTM C227 and C289 have been dropped in both CSA and RILEM as being less reliable in predicting deleterious aggregate performance (Rogers and Hooton, 1989).

While C1260 can act as a rapid screening test, giving results in 16 days, it can be overly aggressive, especially with volcanic rocks. It also cannot be used to evaluate cement-admixture combinations, unless modifications are made to match the storage solution alkalinity to that resulting from the Portland cement used. It can be modified to provide conservative limits on the amount of pozzolan or slag needed to mitigate expansion as has been done in CSA A23.2-27A; and as is currently being considered by ASTM.

ASTM C1293 is considered the most reliable test for both evaluating the reactivity of an aggregate and for selection of mitigative measures. Unfortunately the aggregate acceptance test takes 12 months and to evaluate mitigative measures takes 24 months (at least in CSA).

There are also tests such as ASTM C856 which are used to petrographically examine for the presence of ASR in hardened concretes. AASHTO has also adopted the qualitative uranyl-acetate staining /UV light test.

Table 2
Durability Problem: ASR

Test Method		**Lab-Based**						**Field-Based**	
		C 1260	C 1293	C 289	C 441	C 227	C 295	C 856	Uranyl Acetate
Test Material	Aggregate			x			x		
	Paste								
	Mortar	x			X	x			
	PC Concrete		x					x	x
	SCM Concrete		x					x	x
Properties Measured		Expansion	Exp.	Alkali reduction & silica dissolution	Exp.	Exp.	Mineralogy	Damage, Gel	Reaction Product
Strengths		Rapid, Repeatable							Rapid
Weakness		Not applicable to low alkali cements	Too long	Not a good test	Too reactive & variable	Alkali leaching	Only measures potential risk	After onset	After the fact
Improvements Needed		Reduce false positive	Accelerate						

5. ALKALI-CARBONATE REACTION (ACR)

This type of reactive rock is far less common with reactive deposits in Ontario and Virginia, and possibly in China.

See Table 3. ASTM C1293, or with less effectiveness, the C1105 concrete prism test, can be used to detect ACR expansion. The coarse aggregate needs to be tested in concrete, as crushing the rock to sand sizes, as is done in C227 or C1260 mortar tests, does not allow detection of deleterious aggregates. The C289 quick chemical test is not suitable for evaluation of any carbonate rock types, whether for ASR or ACR. There are no known effective mitigative measures for ACR aggregates (Rogers and Hooton, 1992).

Table 3
Durability Problem: ACR

Test Method		Lab-Based						Field-Based	
		C 1105	C 1293				C 295	C 856	
Test Material	Aggregate						x		
	Paste								
	Mortar								
	PC Concrete	x	x					x	
	SCM Concrete	x	x					x	
Properties Measured		Expansion	Expansion				Mineralogy	Damage, Gel	
Strengths		Repeatable	Faster than C1105						
Weakness		too Slow	Still too long				Only measures potential risk	After onset	
Improvements Needed		Replace with C1293	need to Accelerate						

6. SULFATE RESISTANCE

The deleterious role of C_3A, and the expansive formation of ettringite are the basis for most of the test methods for sulphate resistance. In ASTM C150 there are prescriptive limits on C_3A for Type II ad V Portland cements as well maximum sulfate limits. The sulfate limits can be exceeded, provided that a performance-based mortar bar expansion test, ASTM C1038 is passed (<0.020% after 14 days in water). See Table 4. This is protect against internal sulphate attack. For Type V cements, there is a performance test, C452, to simulate resistance of the cement to external sulfate attack. In this test, the sulfate content of the cement is augmented to 7% and mortar bars are cast. The idea is to have sufficient sulfate next to the C_3A to allow for rapid expansion. A 14-day expansion limit is used to determine acceptance. Since such a test does not allow for the beneficial effects of pozzolans or slag, to be evaluated, ASTM C1012 was developed where mortar bars are exposed to sulfate solutions after a strength of 20 Mpa has been achieved. The downside of this is that the test takes at least 6 months and in some cases 12 months to determine equivalent performance to Type II or V cements, such as in ASTM C1157, C989, C618 and C1240 specifications.

The above tests and limits only deal with the chemical resistance to sulfates and not with preventing or minimising sulphate ingress. The ACI 318 and CSA A23.1 concrete specifications provide limits on W/CM as well as on the Types or performance levels of the cementing materials. There is no standard test method for evaluation of concrete.

There are no standard tests for evaluation of resistance to other forms of sulphate attack, including sulphate salt crystallization, thaumasite, or delayed ettringite formation.

Table 4
Durability Problems: Sulfates

Test Method		Lab-Based					Field-Based	
		C 1038	C265	C452	C 1012		C 876	LPR
Test Material	Aggregate							
	Paste							
	Mortar	x	x	x	x			
	PC Concrete						x	x
	SCM Concrete						x	x
Properties Measured		Expansion	Disolved Sulfate	Expansion	Expansion and Mass loss		Voltage Potential	% Air, Spacing Factor, Specific Surface
Strength		Performance	Semi-performance	Performance	Performance test valid for blended cements and SCM-portland mixtures		Good correlation with f/t performance	Rapid
Weakness		Claims that results not as conservative as C265	Precision of test is poor and is no longer referenced in C150 Spec.	Only valid for portland cements	Takes too long		Sensitive to operator & sample prep. Not a direct measure of f/t.	Not yet in mainstream use
Improvements Needed		Expansion limits and age may need scrutiny			Need to acclerate			

7. FREEZING AND THAWING

It is generally acknowledged that for adequate resistance to cyclic freezing and thawing, the concrete must have sufficient strength prior to freezing, the coarse aggregates must be resistant, and in most cases an adequate entrained air-void system is needed. Specifications such as CSA A23.1, require fresh concrete to have sufficient air, but also require a maximum air-void spacing factor to be achieved in hardened concrete (this is determined using the ASTM C457 microscopic method. Many specifications use tests such as ASTM C88 or the micro-Deval test to determine the frost resistance of coarse aggregates.

a) With Water (See Table 5)
ASTM C666 is used to evaluate the resistance of concrete mixtures to cyclic freezing and thawing while submerged in water (except for Procedure B, when samples are drained prior to freezing). The loss of dynamic modulus of elasticity, due to internal cracking, is used as the measure of resistance. Other cues, such as mass loss or length change can also be used. While this test can be used to show the individual benefits of air-entrainment and frost resistant aggregates, it has been criticised for being overly aggressive (Sturrup, Hooton, Mukherjee, and Carmichael, 1987).

b) With De-Icing Salts (See Table 6)
ASTM C672 qualitatively measures the resistance of concrete surfaces to cyclic freezing and thawing in the presence of de-icer salts. Some agencies in Canada, such as MTO and MTQ have modified this test to include quantitative measurement of the mass of scaled material/unit surface area. One concern with this test is that it appears to be overly severe to pozzolans and slag due to the lack of time prior to exposure for these materials to react.

Table 5
Durability Problems: Freeze-Thaw

Test Method		Lab-Based					Field-Based	
		C 666					C 457	Air Void Analyzer
Test Material	Aggregate							
	Paste							
	Mortar							
	PC Concrete	x					x	x
	SCM Concrete	x					x	x
Properties Measured		Mass loss, Dynamic modulus, Length change					% Air, Spacing Factor, Specific Surface	% Air, Spacing Factor, Specific Surface
Strength		Rapid, Repeatable					Good correlation with f/t performance	Rapid
Weakness		Pre-condition too short for SCM mixtures					Sensitive to operator & sample prep. Not a direct measure of f/t durability.	Not yet in mainstream use
Improvements Needed		Increase maturity of SCM mixtures before test (e.g. VDOT)						

Table 6
Durability Problem: Freezing and De-Icer Scaling

Test Method		Lab-Based					Field-Based	
		C 672	C 672-Mod				C 457	
Test Material	Aggregate							
	Paste							
	Mortar							
	PC Concrete	x	x				x	
	SCM Concrete	x	x				x	
Properties Measured		Visual	Visual & Mass Loss				% Air, Spacing Factor, Specific Surface	
Strength							Good correlation with scaling resistance.	
Weakness		Pre-condition period too short for SCM mixtures	Pre-condition period too short for SCM mixtures				Sensitive to operator & sample prep. Not a direct measure of scaling resistance.	
Improvements Needed		(1) Increase maturity of SCM mixtures before test (e.g. similar to VDOT RCPT). (2) Determine effect of MgCl2	Increase maturity of SCM mixtures before test (e.g. VDOT). Assess results after 5-10 cycles.				Test method to predict field performance in the first year. Correlate field performance to lab test.	

8. CHLORIDE CORROSION OF REINFORCEMENT

Chloride corrosion is a two stage process, where the first defence is to make impermeable concrete with enough depth of cover to extend the time-to-corrosion, and the second stage is to limit the rate of steel corrosion by increasing the resistivity of the pore system. The best way to limit the rate of attack is to reduce the permeability of concrete by specifying low W/CM, use of SCM's, and by proper curing. Other than that, one has to rely on surface coatings which act as barriers between the chlorides and the concrete or by applying barriers on the steel surface.

ASTM C876, the half-cell test (Table 7), is used to determine the probability that corrosion has occurred on the steel directly below the half-cell measurement. It is used to assess the condition of bridge and parking decks.

In the laboratory, the G109 test is used to measure the time to corrosion and later the rate of corrosion of reinforced concrete which is exposed to cyclic wetting and drying with a chloride exposure.
Many of the tests listed in the table (ASTM C1202, C1543, C1556) are for measuring the resistance to chloride penetration of concrete, and not to corrosion per se.

Table 7
Durability Problem: Corrosion

Test Method		Lab-Based						Field-Based	
		C 876	G 109	C 1556	C 1202	C 1543	LPR	C 876	LPR
Test Material	Aggregate								
	Paste								
	Mortar								
	PC Concrete	x	x	x	x	x	x	x	X
	SCM Concrete	x	x	x	x	x	x	x	X
Properties Measured		Voltage Potential	Time-to corrosion and rate of corrosion	Bulk Diffusion	Conductivity			Voltage Potential	% Air, Spacing Factor, Specific Surface
Strength								Good correlation with f/t performance	Rapid
Weakness								Sensitive to operator & sample prep. Not a direct measure of f/t.	Not yet in mainstream use
Improvements Needed									

9. ABRASION RESISTANCE

Resistance to abrasion is achieved by deign of low W/CM, well-cured concrete made with abrasion resistant aggregates. Additional benefits can be attained through use of silica fume (to improve bond of paste to aggregates) and by use of fiber reinforcement (to reduce loss of surface). There are several ASTM test methods (Table 8) contained in C994, C131, C535, C779, and the micro-deval test (CSA and Europe). All of these tests measure mass loss of a surface. The micro-deval test is used to measure abrasion resistance of aggregates.

Table 8
Durability Problem: Abrasion

Test Method		Lab-Based						Field-Based	
		C 944	C 131	C 535	C 779	Micro Deval			
Test Material	Aggregate								
	Paste								
	Mortar								
	PC Concrete				x				
	SCM Concrete				x				
Properties Measured		Mass loss	Mass loss	Mass loss	Mass loss	Mass loss			
Strength									
Weakness		Sensitive to specimen prep. Not applicable to textured surfaces.	Does not test paste, mortar or concrete	Does not test paste, mortar or concrete		Does not test paste, mortar or concrete			
Improvements Needed		Reproducibility							

10. ACID RESISTANCE

Acids will reduce the alkalinity of the pore solutions and result in destabilisation and dissolution of C-S-H as well as calcium hydroxide. The best way to limit the rate of attack is to reduce the permeability of concrete by specifying low W/CM, use of SCM's, and by proper curing. Other than that, one has to rely on surface coatings which act as barriers between the acid and the concrete.

11. RESISTANCE TO VOLUME CHANGE AND CRACKING

Cracking of concrete due to plastic shrinkage, autogenous shrinkage, drying shrinkage, thermal shrinkage, and creep are of concern to durability of structures since cracks act as rapid pathways for penetration of water and other aggressive fluids into concrete. The effects of cracking are to accelerate degradation. Table 9 lists some of the tests currently available to measure free (C157) and restrained (AASHTO Ring Test) drying shrinkage. The ASTM fiber subcommittee is currently developing a similar restrained shrinkage ring test and two plastic shrinkage test methods. ASTM subcommittee C01.31 is currently considering a method to measure chemical shrinkage of cement paste. ASTM has a creep test, C512, and the RILEM early-age volume change committee is working on a tensile creep apparatus to measure early age deformations due to various early-age stresses.

The Corps of Engineers has standard methods used to determine thermal properties of concrete of concrete, including thermal coefficient of expansion, specific heat and heat capacity.

Table 9
Durability Risk: Volume Change/Cracking

Test Method		Lab-Based ASTM C 157 Drying Shrinkage	AASHTO Ring Test	RILEM Proposed Tensile Creep	ASTM C 512 Creep	Specific Heat	Thermal Coefficient of Expansion
Test Material	Aggregate						
	Cement					x	
	Paste						
	Mortar	x					
	PC Concrete	x	x	x	x		
	SCM Concrete	x	x	x	x		
Properties Measured		Length Change. Length Change vs Time Curve.	Cracking-visual	Length Change	Length Change. Length Change vs Time Curve.	Energy-Input & Output	
Strength							
Weakness		Too long. Operator sensitive. Lack of correlation to field performance.			Too long. Operator sensitive. Lack of correlation to field performance.	Lack of correlation to field performance	
Improvements Needed		Shorten duration. Establish relationship to field conditions (restraint, etc.) and incidence of cracking.	Establish relationship to field conditions (restraint, etc.) and incidence of cracking.	Establish relationship to field conditions (restraint, etc.) and incidence of cracking.	Shorten duration. Establish relationship to field conditions (restraint, etc.) and incidence of cracking.	Establish relationship to field conditions (restraint, etc.) and incidence of cracking.	

12. LACK OF SKILLED OR CERTIFIED LABOR

While engineers require professional licencing, most designers have structural or construction management experience, and few have extensive education or knowledge in concrete materials or in design durable concrete.

Contractors are only required to be able to obtain a bid bond, that is only limited in size by the size of their previous one. There is no formal knowledge of concrete durability required.

The construction labor force includes various skilled trades people as well as semi-skilled laborers. Because of the short-term nature of construction, many of these trades people and laborers are only hired for the period of construction, and therefore are transient.

The ready-mixed concrete truck driver has great control over the in place quality of concrete, since he/she controls the amount of water added on site. The only qualification of most drivers is a driver's licence.

Until something is done to raise the level of concrete education and requirements for certification of concrete workers are put in place, concrete will continue to suffer premature cracking or degradation due to the lack of a concrete–skilled labor force. Anything that can be done to train or educate the work force, especially construction laborers, and supplier industry staff, will likely have a bigger impact on performance of concrete than all the new test methods that can be developed.

13. CONCLUSIONS

It is clear that in order to change current prescriptive specifications towards performance specifications, new test methods are needed and improvements are needed to existing test methods, to better address all of the performance issues and provide confidence to specifiers.

REFERENCES

El-Dieb, A.S. and Hooton, R.D., "Water Permeability Measurement of High Performance Concrete Using a High Pressure Triaxial Cell", Cement and Concrete Research, Vol. 25, No. 6, 1995, pp. 1199-1208.

Julio-Betancourt, G.A. and Hooton, R. D., "Study of the Joule Effect on RCPT Values and Evaluation of Related Electrical Properties of Concrete", Cement and Concrete Research, MS#5397, accepted, Sept. 6 2003.

McGrath, P.F. and Hooton, R.D., "A Re-Evaluation of the AASHTO T259-90 day Salt Ponding Test", Cement and Concrete Research, Vol. 29, 1999, pp. 1239-1248.

Rogers, C.A. and Hooton, R.D., "Concrete and Mortar Expansion Tests for Alkali-Aggregate Reaction - Variations Due to Alkali Leaching", Proceedings, Eighth International Conference on Alkali-aggregate Reaction, Kyoto, July 17-20, 1989, pp. 327-332.

Rogers, C.A. and Hooton, R.D., "The Relationship Between Laboratory and Field Expansion of Concrete Affected by the Alkali-Carbonate Reaction", Proceedings, 9th International Conference on Alkali-Aggregate Reaction in Concrete, Concrete Society, London, July 27-31, 1992, Vol. 2, pp. 877-884.

Roumain, J. C., 2002. "Standards and Concrete Durability: An Industrial View" Designing For concrete Durability, Anna Maria Workshop, Florida. (Powerpoint presentation).

Stanish, K., Hooton, R.D., and Thomas, M.D.A., A Rapid Migration Test for Evaluation of the Chloride Penetration Resistance of High Performance Concrete, in: P. Johal, (Ed.) Proceedings of the PCI/FHWA/fib Symposium on High Performance Concrete, Sept. 25-27, 2000, Orlando, 2000, pp. 358-367.

Sturrup, V.R., Hooton, R.D., Mukherjee, P.K. and Carmichael, T.J., "Evaluation and Prediction of Concrete Durability - Ontario Hydro's Experience", Proceedings, Katherine & Bryant Mather International Conference on Concrete Durability, Apr. 27-May 1, 1987, Atlanta, Ga., 34 pp., ACI SP-100, Vol. 2, pp.1121-1154.

Tang, L., and Nilsson, L-O., Rapid Determination of the Chloride Diffusivity in Concrete by Applying an Electrical Field, ACI Matl J, 89(1) 1991, 49-53.

Thomas, M.D.A., Hooton, R.D. and Rogers, C.A., "Prevention of Damage Due to Alkali-Aggregate Reaction (AAR) in Concrete Construction - Canadian Approach", Cement, Concrete and Aggregates, Vol. 19, No. 1, June, 1997, pp. 26-30.

ANNA MARIA WORKSHOP 2003

TESTING & STANDARDS FOR CONCRETE DURABILITY

WORKING GROUP 3

MATERIALS SELECTION AND PROPORTIONING FOR DURABILITY

Group members:

Sidney Mindess (co-chair)	University of British Columbia, Canada
Jeff O'Leary (co-chair)	Florida Rock Industries, USA
Arnon Bentur	Technion, Israel
Sharon DeHayes	Rinker Materials, USA
Kevin Folliard	University of Texas, USA
Juraj Gebauer	Holcim Group Support, Switzerland
Jan Olek	Purdue University, USA
Jan Skalny	Materials Service Life, USA
Niels Thaulow	RJ Lee Group, USA
Saneel Vanikar	Federal Highway Administration, USA

1. INTRODUCTION

The focus of the Anna Maria 4 Workshop was:

1. Identification of needs and opportunities for improving the transfer of knowledge required in the formulation, production, and use of concrete for which service life ("durability") can be reliably predicted;
2. Highlighting outstanding durability issues as they relate to standards and specifications; and
3. Drafting of recommendations for steps to be taken to bring about the needed improvements in knowledge transfer.

Within this framework, the objective of Working Group 3 was ***to establish criteria and procedures for the selection and proportioning of the materials for a concrete mixture to ensure durability in field applications.*** Durability itself may be defined as the ability of the concrete to perform satisfactorily, with minimal maintenance, for the anticipated service life of the project. Ideally, the performance of concrete placed in an aggressive environment should be similar to that of concrete placed in a non-aggressive environment. Unfortunately, however, we generally are unable to **quantify** the concept of "service life". That is, we cannot accurately **predict** the effective life of the concrete in a structure based only upon knowledge of the exposure conditions and the concrete constituents. We simply assume that, if we follow current guidelines properly, we can anticipate a service life of at least 50 – 100 years.

Of course, **we do know how** to make durable concrete for most service conditions, and much of the concrete that we produce currently is of high quality. In spite of this, however, a great deal of poor quality, non-durable concrete is still being placed, as evidenced by the rapid growth in construction litigation related to concrete. Clearly, then, the present system does not work as well as it should; changes in how we specify concrete, and how we ensure concrete quality, are warranted. We must therefore devise some practical steps toward a more rational use of *all* of the concrete-making materials (and know-how) that are readily available. The challenges that we now face are threefold:

1. How do we transfer our knowledge to the field?
2. How do we select reliable test methods and criteria for durability?
3. How do we ensure compliance with the durability criteria?

2. BASIC PRINCIPLES

In keeping with the mission to incorporate service life requirements (and the means and motivation to enforce them) into design, there are three basic principles:

1. It is essential that requirements for durability (service life) as they exist in current codes and specifications be integrated into all phases of specification, design, construction and QA/QC.
2. It should be the responsibility of the **owner** to ensure that durability issues are properly addressed.
3. An "**engineer of record**" should be accountable for the inclusion of the appropriate durability considerations in the design and specifications, and for insuring that these considerations are implemented.

While adherence to these principles will not *guarantee* concrete quality, they will at least ensure that durability is considered "up front" in the design process. The cost of producing durable concrete is not excessive. For instance, to go from a *w/cm* ratio of 0.7 to a *w/cm* ratio of 0.45 would cost only about US\$13.00/yd^3 (based on 2002 prices in Ontario, Canada).

3. FROM PRESCRIPTIVE TO PERFORMANCE SPECIFICATIONS

In recent years, we have made enormous advances in the types of concretes that we can produce, including *very high strength* concretes, *self-consolidating* concretes, and *tough* concretes, and in our ability to "tailor make" concretes for a wide range of special applications. However, in producing these concretes, and even more so for the "normal" concretes that make up the bulk of our production, we still rely almost entirely on **prescriptive** specifications. Such specifications generally include requirements such as maximum w/cm ratios, minimum cement contents, cement types, limitations on the types and amounts of both chemical admixtures and mineral admixtures, and so on. While these types of specifications have served us reasonably well in the past, when the industry as a whole was much less sophisticated than it is now, they also tend to inhibit the most efficient use of the materials now available to make up a concrete mixture. And, as noted above, they are far from a guarantee of concrete durability. Thus, it is now time for us to move to a much more extensive use of **performance based** specifications. If these can be properly devised, they would permit concrete producers to be more imaginative and innovative in their use of materials, such as supplementary cementing materials, admixtures, blended cements, polymers, fibers, mineral fillers, etc. They would also provide a means for introducing durability concerns more explicitly into the design of concrete mixtures.

There are, of course, a number of obstacles to the introduction of performance based specifications:

- Concrete producers, and in particular the smaller producers, are not really prepared to make the switch from prescriptive to performance based specifications. There is often a lack of properly trained personnel to handle the issues that would arise regarding quality control, and to provide the necessary technical advice to specifiers.
- There is a lack of quick, reliable tests for concrete durability, which would have to go far beyond our reliance on the 28-day compressive strength as the sole arbiter of concrete quality. This lack of adequate performance-related test methods is one of the main factors inhibiting the move from prescriptive to performance specifications.
- There would have to be a procedure in place for **assigning responsibility** for adequate concrete design. That is, there must be some way of giving the consumers (buyers) of the concrete some confidence that the concrete is indeed suitable for its intended purpose. Currently, this responsibility is rather diffusely shared amongst the geotechnical engineer, the structural engineer, the cement producer, the concrete supplier, the contractor, the developer, and perhaps others as well. This is not a situation designed to inspire confidence in the purchaser.

The challenge for the industry is to devise a road map for moving from prescriptive to performance based specifications that would, over time, lead to a more rational and efficient use of all of the concrete-making materials that are now available, or that will become available in the future.

4. GETTING FROM HERE TO THERE

In order to move from primarily prescriptive specifications to primarily performance based specifications, we have a number of proposals. Collectively, these would effectively "raise the bar" for concrete production:

1. All concrete producers should be **certified** (by either state or federal organizations or agencies). This would require *legislation*, similar to that governing other professional organizations.
2. The addition of mixing water on site (beyond that required to bring the concrete mix up to the maximum specified w/cm ratio) should be absolutely forbidden. This too would require a legislated change to building codes.
3. Curing procedures must be specified and monitored.
4. Durability should not be defined or measured or in terms of compressive strength, since the two are not directly related. This would require (alas) both massive re-education of engineers/technologists/specifiers/suppliers, and some changes to building codes.
5. Durability should be measured on the concrete system, rather than on the individual concrete constituents. In some cases, such as for sulfate resistance, tests only exist which

attempt to evaluate the chemical resistance of the cementing system (*e.g.*, ASTM C452, ASTM C1012) but not the concrete quality (*e.g.*, *w/cm limits).* Typically, concrete quality limits are contained in specifications such as ACI 318. (While these could be considered to be prescriptive, they are still a necessary part of current specifications). As stated above, this would first require the development of quick and economical test methods for long-term concrete properties (durability), including those for sulfate resistance, freeze-thaw resistance, alkali-aggregate reactivity, marine exposure, exposure to other aggressive chemicals, and so on. Some suitable test methods may already exist in ASTM and other national or international standards, but these methods do not address all of the durability concerns. In addition, some of these methods are outdated, and have not kept up with the state-of-the-art technology and instrumentation. Others do not correlate well with field experience.

6. We must also develop methods for the determination of the w/cm ratio in both fresh and hardened concrete, for concrete permeability in both laboratory and *in situ* concrete, and for the moisture content *in situ* of the concrete.
7. The issue of the proper handling and placing of the concrete must be addressed in conjunction with the materials considerations.
8. There must be more (and better) education in concrete science and technology for engineers, specifiers, producers, contractors, etc. The universities, the technical colleges and the professional and trade organizations all have a role to play in this regard (a role which none of them has fulfilled very well in the past).
9. At least some of the funding for "academic" cement and concrete research should be re-directed to technology transfer. As stated at the outset, we already know how to produce durable concrete – we just have to make sure that this knowledge is properly transmitted to field practice. The various federal and state agencies should collectively try to develop a research agenda which will focus funding on the most important durability issues.
10. A "check list" of durability issues should be created, for mandatory use by specifiers, contractors, and producers, to help ensure that durability issues are not overlooked in the process of specifying and producing the concrete. An example is shown in Table 1.
11. A matrix of supplementary cementing materials should be created, which would address their application to various aggressive environmental conditions. A similar matrix of readily available chemical admixtures should also be available.

Table 1
CONCRETE SPECIFIERS/PRODUCERS CHECKLIST

	Yes	No	If yes
Application Location Exposure: - freeze-thaw - sulfates - de-icer salt scaling - steel corrosion - ASR/ACR - acids - other aggressive chemicals - seawater - carbonation [CO/CO_2] - plastic shrinkage -crack resistance -architectural			 Air entrainment Low w/c; Type II or Type V cement
Mixture Minimum specified strength Nominal maximum aggregate size Slump range Air Any other special requirements Quantity Required Rate of Delivery			 E.g; concrete set time requirement
Transport and Placement method From chute Pump Conveyors			
Curing Method			

5. INTERIM MEASURES

Clearly, the transition from prescriptive to performance based specifications, in conjunction with the implementation of the proposals listed above, will take a relatively long time. We must therefore adopt some interim measures during this transition, and we must also recognize that we may never be able to move away completely from some form of prescriptive specifications. Thus, until the proposals above are implemented, we will still have to rely heavily on w/cm ratio requirements (though not $f_{c)}$ to ensure durability. This could take the form of an extensive table giving maximum w/cm ratios for a variety of exposure conditions. Also, for small producers and/or small projects, it would be more efficient to maintain some prescriptive specifications, as long as these are sufficiently conservative.

In addition, while we wait for the necessary test methods to be developed, we may have to rely upon the "equivalent performance" concept. That is, a new concrete mix will be deemed to be satisfactory as long as it behaves at least as well as a mix already known to be durable under the expected service conditions.

6. CONCLUSIONS

The above recommendations do not, of course, address all of the problems associated with producing durable concrete. In particular, the issues of **how much** durability we really need, and **who pays** for it, have not been discussed. The question of whether durability can be "profitable" (making durable concrete at the outset is almost certainly always less costly than litigation) is also worthy of consideration. Nonetheless, the modest proposals made above could go a long way towards ensuring the viability and growth of the concrete construction industry.

ANNA MARIA WORKSHOP 2003

TESTING & STANDARDS FOR CONCRETE DURABILITY

WORKING GROUP 4

STRUCTURAL DESIGN AND DETAILING CONSIDERATIONS TO ACHIEVE DURABILITY

Group Members:

Denis Mitchell (chair)	McGill University, Canada
Jim Pierce (co-chair)	Bureau of Reclamation, USA
Andrzej Brandt	Polish Academy of Sciences, Poland
Stephen Forster	FHWA, USA
Rich Lee	RJ Lee Group, USA
Marjorie Lynch	Surtreat NE, USA
Donald Meinheit	Wiss, Janney & Elstner Associates, USA
Ken Rear	Heidelberg Cement, USA
Paul Tourney	Materials Service Life, USA
Carl Walker	Carl Walker Construction Group Inc., USA

The purpose of this report is to provide a guide for structural engineers in the design and detailing of concrete structures to achieve the desired level of durability for a particular project.

1. STEPS REQUIRED TO ACHIEVE DURABILITY

Figure 1 shows the steps required for the design and detailing of concrete structures to achieve durability. These steps are discussed in more detail in the report below.

2. STRUCTURAL TYPES, SERVICE LIFE EXPECTATIONS AND EXPOSURE CONDITIONS

Three aspects that have a major bearing on the considerations for achieving durability for a particular project are: the type of structure; the expected service life; and the exposure conditions. These three aspects must be considered together in assessing the necessary measures to achieve the desired level of durability.

Structural types that need special attention include parking structures, marine structures, pavement, underground structures (e.g., tunnels, sewer systems) and foundations.

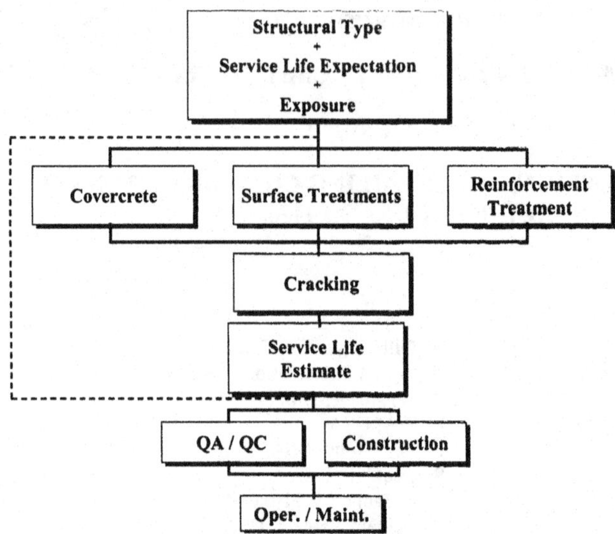

Figure 1 – Steps in the design and detailing of concrete structures to achieve durability

The expected service life for projects has become an important issue for the client or agency controlling the project. It is prudent that the structural designer discuss the expected service life with the client at the beginning of a project so that design and detailing considerations affecting durability can be made. It is important that the client be made aware of the choices available and the likely effects of these choices on the first cost and the likely savings associated with extending the service life, including maintenance and repair costs over the life of the structure.

The exposure conditions depend on the natural environmental and the environment imposed on the structure due to its use. ACI Committee 362 (ACI 1997 identifies 3 climatic zones and a coastal zone for assessing the environmental exposure. Climatic zones, affecting design and detailing of bridge structures are given by the AASHTO Specifications (AASHTO 2000). ASTM C33 (ASTM 1999) identifies freeze-thaw zones depending on the severity of freeze-thaw cycling. A key feature is whether or not the structure will be subjected to chlorides from de-icing salts, ground chlorides or coastal salts. Aspects such as exposure to water (especially water ponding), a significant number of freeze-thaw cycles, high CO or CO_2 environments and sulfate exposure all have a detrimental affect on durability. Cyclic thermal effects (e.g., bridge decks and roofs of parking garage structures) cause cyclic tensile stresses that can reduce the durability.

3. COVERCRETE

One of the most important features that influences the durability is the quality and thickness of the concrete cover or "covercrete" over the reinforcement. Key aspects of the covercrete that affect the durability are its permeability, the curing regime, the tolerances of construction, any surface treatment and the presence of chemical admixtures. The presence of cracks in the covercrete significantly reduces the durability. Other factors that can reduce the durability are abrasion, freeze-thaw effects, chemical attack, etc. The thickness and quality of concrete are key factors in providing fire resistance and bond strength between the concrete and the reinforcing bars. Cover deterioration and corrosion of the reinforcement often leads to bond failure.

Requirements for the cover concrete and minimum clear cover dimensions for structures subjected to severe exposure conditions (parking structures) are given by ACI Committee 362 (ACI 1997), the Precast/Prestressed Concrete Institute (PCI 1997) and the Canadian Standards Association Standard S413 (CSA 1994). It is noted that because of the better control of tolerances and the improved control on the concrete properties in precast concrete construction, precast elements typically have smaller minimum cover requirements for the same exposure conditions than cast-in-place structures (ACI 2002, CSA 2000).

4. SURFACE, CONCRETE TREATMENTS AND DRAINAGE

Surface treatments for concrete can be divided into two categories, surface treatments and concrete treatments. Surface treatments include overlays, sealers (including penetrating sealers) and membranes. Membranes include traffics and non-traffic bearing membranes. It is noted that all surface treatments need maintenance.

The use of supplementary cementing materials such as silica fume, fly ash, slag, metakaolin, ternary blends and quaternary blends typically result in lower permeability and longer service life. Chemical admixtures that enhance the durability include water-proofing admixtures, high-range water reducers, chemical corrosion solutions, air-entraining agents and nucleation agents that aid hydration. Other solutions involve the use of synthetic fibers for the control of micro-cracking, structural fibers to control service load cracking and impregnation (e.g., epoxy in thin overlays).

A key feature for achieving durability in structures is to provide adequate drainage. Adequate slopes must be provided for drainage and the design of these slopes must account for cambers and/or deflections of the structural elements. In designing drainage systems it is important, particularly for precast concrete structures, to drain away from joints. The drains must also be adequately sized and must allow for cleaning, clogging and freezing.

5. REINFORCEMENT TREATMENTS

A variety of protection systems for reinforcement are available. These include stainless steel reinforcing bars, coated stainless steel bars, MMFX reinforcement, fiber-reinforced plastics

(FRP) bars, galvanized reinforcing bars and epoxy-coated reinforcement. Cathodic protection may also be used for corrosion protection.

6. CONCRETE CRACKING

Cracking of the concrete cover can play a significant role in reducing the long-term durability of concrete structures. Early-age cracking can result from thermal gradients, drying surface shrinkage, plastic shrinkage and restraint effects that cause tensile stresses. The type and timing of curing is critical to avoid early-age cracking, particularly for high-performance concrete. During hydration the material properties (modulus, tensile strength and early-age creep) are changing as the tensile stresses develop. Restraint effects are caused by substrates and supporting members.

Cracking during service can also compromise the durability. In-service cracking can occur due to cyclic thermal changes, normal load effects and differential settlements. Overloads, for example, overloads on pavements and bridges as well as fatigue and impact effects can result in significant cracking under service conditions.

Expansion joints can be used effectively to relieve volume changes such as shrinkage, creep and temperature change. Care must be taken to account for leakage of expansion joints and to account for load transfer across the joints and to minimize local damage across the joints. The spacing, size, location and sealing of the expansion joints must be carefully designed and detailed to minimize restraint effects and ensure long-term durability. Expansion joints must also be well maintained to remain effective.

Construction joints and pour strips must also be carefully located in order to minimize restraint effects. Often a joint sealant (groove detail) will be used or a membrane will be placed over the joints. Pavements typically have dowel bars that must be properly aligned to minimize restraint effects, with the bars typically epoxy coated or stainless clad to improve durability. Joint keys may require special detailing and leak-proof structures will typically require water stops.

The choice of reinforcing types and details can have a significant effect on the structural durability. Prestressed concrete construction offers the benefit of controlling the tensile stresses in the precompressed tensile zone. The details of the reinforcement are also very important in terms of the durability of a structure. Large diameter reinforcing bars with small concrete covers results in poorer control of cracking than more closely spaced smaller bars. However, the clear spacing between bars affects the constructability in that small spacings may not permit the proper consolidation of concrete around the bars. Increased use of headed reinforcing bars in lieu of hooked anchorages provides better bond and anchorage characteristics and less cracking at service load levels. Other options to consider involve the replacement of bar splices with couplers. In slab-like elements, especially those subjected to corrosive environments, the traditional minimum amount of reinforcement (temperature and shrinkage reinforcement) is not sufficient for ensuring long-term durability.

7. SERVICE LIFE ESTIMATES

There is an increasing need to provide realistic estimates of the service life of reinforced concrete structures. A number of service life prediction models are either available or under development. Important questions in selecting a service life prediction model include the following:

- Is the exposure condition that you wish to model properly addressed by the model?
- Has the model been adequately validated in terms of numerical validation, laboratory validation and field validation?
- Is cracking of the concrete accounted for?

In any case the designer should perform a sensitivity study to see the predicted influence of increased cover, concrete properties, curing and exposure conditions to appreciate the capabilities of the model and some of the major factors affecting durability.

As with many computer simulations one must realize the reliability of the predictions and the importance of carefully choosing the key parameters to achieve sensible results.

After the designer has made predictions of the service life, it may be necessary to return to some of the design parameters if the predicted service life is deemed inadequate (see dashed lines in Fig. 1).

8. QUALITY ASSURANCE AND QUALITY CONTROL (QA AND QC)

The quality assurance and quality control must be addressed at both the design stage and the construction stage.

During the design stage, the service life expectations should be clarified at the beginning of the job. The designer should consider a "durability checklist", such as that illustrated in Fig. 1. For severe environments the structural designer should consult durability experts to obtain the latest information and to provide realistic estimates of the service life. The designer should provide an operations/maintenance plan for the owner of the structure.

During the construction stage, a series of quality control checks should be made on-site, including:

- measurements of the locations of the reinforcement before concreting,
- measurements of the positions of screed rails,
- periodic cover measurements after concreting,
- the type of finishing,
- the duration and type of curing,
- the qualifications (certification) of the work force, and
- the results of concrete performance tests.

The variances in the assumptions made in the original design can be evaluated using appropriate service life models.

9. CONSTRUCTABILITY AND CONSTRUCTION

Successful construction requires coordination, logistics and teamwork. It is important to lay the groundwork for these three important aspects at the beginning of a project. The use of construction incentives, including penalties and rewards, together with effective control measurements (see above) can be an effective in achieving more durable concrete structures.

10. OPERATION AND MAINTENANCE

There is a trend towards Design, Build, Operate and Maintain (DBOM). This approach spreads the responsibility over the life of the structure and provides the greatest incentive for achieving durable structures. Other approaches involve construction incentives, as described above and also the use of "extended warranties". ACI Committee 362 has provided a guide for the structural maintenance of parking structures (ACI 2000).

REFERENCES

AASHTO, 2000, "AASHTO LRFD Bridge Design Specifications", Second Edition, 2000 Interim, American Association of State Highway and Transportation Officials, Washington, DC.

ACI, 1997, "Guide for the Design of Durable Parking Structures (ACI 362.1R-97)", ACI Committee 362 – Parking Structures, American Concrete Institute, Farmington Hills, MI.

ACI, 2002, Building Code Requirements for Structural Concrete (ACI 318-02) and Commentary (ACI 318R-02), American Concrete Institute, Farmington Hills, MI.

ACI, 2000, "Guide for Structural Maintenance of Parking Structures (ACI 362.2R-00)", ACI Committee 362 – Parking Structures, American Concrete Institute, Farmington Hills, MI.

ASTM, 1999, "ASTM C33-99a – Standard Specifications for Concrete Aggregates", American Society for Testing Materials, West Conshohocken, PA.

CSA, 1994, "Parking Structures, CSA S413-94", Canadian Standards Association, Rexdale, Ontario.

CSA, 2000, "Concrete Materials and Methods of Concrete Construction/Methods of Test for Concrete (CAN/CSA-A23.1-00, CAN/CSA-A23.2-00), Canadian Standards Association, Rexdale, Ontario.

PCI, 1997, "Parking Structures: Recommended Practice for Design and Construction, Precast/Prestressed Concrete Institute, Chicago, IL.

III. Papers Presented at the 5th Anna Maria Workshop (2004)

100 YEARS OF CEMENT AND CONCRETE
Successes and Challenges

Juraj Gebauer
HOLCIM, Switzerland

Jan Skalny
Materials Service Life, USA

ABSTRACT
For about 100 years now portland cement-based concrete contributed to shaping the industrial revolution experienced during our life-times. Without concrete, there wouldn't be modern transportation, large metropolitan areas wouldn't exist, the rate of increase in the level of living would be dramatically lower, and the global industrial expansion would have been marginal. A brief review of the historical literature shows great successes, but also missed opportunities – both in research and in practical applications.

After a brief review of the most important successes of cement-concrete research and technology, we will attempt to openly discuss some controversial technical and institutional challenges that hinder more intelligent, economic, ethical, and ecologically sound utilization of concrete. A few recommendations will follow.

1. INTRODUCTION

One does not have the experience and credibility to be outspoken when young, one cannot risk his career/job when he's middle-aged, and most of us are skeptical or apathetic to expose our views when we supposedly (!?) mature. It is however a liberating emotion to express yourself freely and to tell your friends and colleagues how you see your professional world without pretenses, political games, or anger. We believe this is our time to do it and this is the forum to do it; however, we realize we're preaching to an already converted audience and our views will not make a lasting impact on those who should, but don't, listen.

Both of us were for almost 50 years a part of the cement-concrete materials community, exposed to the excitements and fallacies of the research environment, but also to the realities of marketing, production, and corporate politics. This experience was exciting, controversial, mostly based on honest disagreements, sometimes nasty; it quite often increased one's adrenaline level and it led to temporary frustrations. Was this experience different from other aspects of life? Not at all! What's going on in research and corporate communities reflects the character of humans in general; nobody is only good or bad; we're all human, thus imperfect!

2. PAST PROGRESS AND SUCCESSES

Portland cement-based concrete is one of the building blocks of modern industrial world, and the science and technology that resulted in its wide use is an amazingly complex system. We won't discuss all the scientific advances in detail, as many excellent papers on the topic were published or presented by others [e.g., 1,2]. Just let's remember that the cement and concrete technology successes are based on a multitude of scientific disciplines, including geology, mineralogy, chemistry, physical chemistry, physics, process engineering, soil mechanics, hydrology, fracture mechanics, etc. – a most complex system, not easily manageable on an industrial scale.

The Cement Industry

The development of Portland clinker based cements is – notwithstanding some justified criticism – a great and lasting success story [3]. An enormous progress has been realized primarily during the last century, when production of portland clinker-based cement rose by a factor of 200, from 10 million tons to present 2,000 million tons per year, representing a dollar value of about $200 (U.S.) billion. Assuming that the cost of cement is about $100/ton and that in an average structure the cement represents about 2% of the overall cost, this amount represents an inconceivable value of about $10,000 billion.

Thanks to the pioneering inventions during the 19th and 20th Centuries - including those by Smeaton, Aspdin, Le Chatelier, Michaelis, Edison, Bogue, and many others – the production of portland and blended cements experienced a tremendous growth, allowing establishment of a new successful industry, one of the key industries responsible for the recent extraordinary technological progress of mankind. What were the primary achievements of the industry during the past 100 years? In our view, most of the progress was recorded in the manufacturing area!

Clinker Production: Clinker manufacturing progressed from small-sized vertical shaft kilns to wet rotary kilns, to semi-wet (Lepol) kilns, to dry kilns with suspension preheaters and precalciners. The modern kilns of today are short dry kilns with up to 6-stage suspension preheaters and a precalciner. The thermal energy consumption decreased from about 10,000 kJ/kg of clinker for the long wet kilns at the beginning of the 20th century to about 3,000 kJ/kg of clinker for the modern dry kilns with suspension preheaters and precalciner.
The increase of the output of kilns during the past 100 years is also impressive: from about 30 tons to up to 10,000 tons per day!! The new tools in automation and high-level control facilitated the introduction of the energy saving large units. For more than 30 years, the cement industry has been exploring computer-based techniques to control and optimize the operation of cement kilns. The major reasons for introduction of these techniques were clinker uniformity, savings in energy consumption, increase in production volume, longer refractory life, and reduced NO_x emissions. F.L.Smidth supplied the first kiln control system based on fuzzy logic in 1980. Since that time, many new systems have been developed and marketed. One of the most successful is the ABB "Linkman" system, a powerful real time expert system.

During the past 20 years, significant developments took place in the use of alternative fuels leading to numerous environmental benefits. In contemporary modern kilns up to 70 % of fossil fuels can be replaced by alternative waste fuels; on average, bout 15% replacement is realized already.

Application of computer science enabled great improvement in the exploitation of raw materials. Computer aided evaluation of raw materials availability, quarry scheduling optimization, and

quarry engineering/design are some of the tools for the optimal utilization of raw material resources and for helping to assure uniformity of the kiln feed.
Since 1990 the production of mineralized clinker becomes again important, using fluxing agents, such as fluorides and sulfates, to improve clinker reactivity and thus enable production of cements with higher quantity of mineral components.

***Cement Manufacturing*:** Here we would like to first mention the progress made in the grinding technology. The simple open circuit ball mill advanced to sophisticated closed circuit mills with high efficiency separators, to roller press technology for pre-grinding and finish grinding, to horomill (horizontal roller mill), and vertical roller mill. Today the vertical roller mill is routinely used for grinding raw materials; its application for cement grinding seems to have gained importance only within the last couple of years, after solving some important process and product quality problems. With the introduction of modern grinding technologies, energy saving (costs of wear and maintenance) of about 40 % are possible compared to the classical ball mill grinding. For finish grinding of portland cement to a specific surface area (Blaine fineness) of 350 m^2/kg as little as 25 kWh/t are required in a modern cement mill. Thanks to new technologies, the production rate increased from about 10t/h at the beginning of the 20th century to up to 200t/h for modern mill units.

Slow but steady increase of the production of portland blended cements took place in the 20th century, from the first slag cement produced in Europe in small quantities to great number of different types of portland blended cements using blast furnace slag, natural pozzolans, fly ash, silica fume, limestone, and other mineral components. Today about half of all cements produced are portland-clinker based blended cement with an amount of more than 5% of mineral components. The so-called clinker factor also gradually improved worldwide, on average from 90-95 to 70-80. There is still a potential for further improvement in this respect. The use of mineral components in portland cement enables, beside environmental and economical advantages, production of cements with better durability and resistance to chemical attack.

The automation of quality control in the cement plant experienced a significant technical progress too, mainly during the past 30 years. Quality/uniformity improvements and cost optimization of large production units by means of online bulk material analyzers on belt or stockpile, as well as online cement fineness measurement by automated Blaine instrumentation or laser granulometry, are routinely used today. Fewer changes were realized in testing of the finished product. The testing methods established more than 100 years ago are, in principle, still used today (setting time, compressive strength testing on mortar, etc.). There is an effort to introduce modern automated testing method at least for the internal quality control of the finished product. The main reason for the slow development in this area is the conservatism of the relevant standardization committees.

***The Finished Product*:** In contrast to the progress in the manufacturing process, there was little change or improvement realized in the chemistry and properties of clinker and cements. The gradual increases of lime saturation, alite and alkali content of clinker, and fineness of the cement during the past 100 years have led to production of cements with higher chemical reactivity and early strength - not necessarily a positive development with respect to concrete performance, especially regarding chemical durability.

The number of cement types produced increased somewhat, mainly due to the use of mineral component, but principally there were no meaningful changes or new developments in the chemistry of cement hydration. Beside the portland clinker-based ordinary and blended cements, special cements such as masonry cement, oil-well cement, white cement, refractory cement (HAC), super-sulfated cement, expansive/shrinkage compensating cement, fast setting cement, and alkali activated cement were developed and established on the market with different degrees of success and acceptance.

Most of the above mentioned special cements are based on portland cement or modified portland cement clinker [4]. The few non-portland cements, such as calcium sulfoaluminate-belite cements, were developed and marketed as special cements for a particular application. The common important features of such non-portland cements are primarily the lower calcium content and lesser water requirement compared to ordinary portland cement, very rapid strength development and good resistance against chemical attack. This is caused mainly by the presence of strength-producing ettringite and the absence of calcium hydroxide in the hydrated/hardened cement paste. Another advantage of some of the non-portland cements is their better environmental compatibility (lesser CO_2 emissions during the clinkering process) compared to portland cement.

Numerous attempts to replace portland cement in common concrete have failed thus far, mainly due to cost and availability of raw materials, manufacturing cost, and some deficiencies in performance. The strong protection of the traditional cements by the cement industry, the conservative standards and certification procedures, difficulties in market acceptance, and the very high R&D cost - in particular in the scale-up phase (industrial production and commercialization) - are additional reasons for the failures.

Concluding Remarks: The highlighted improvements in the manufacturing process are the primary reasons for increased efficiency and economy of the cement production. If we compare the man-hours needed to produce a ton of clinker, we have made a very substantial progress: today we need on average about 0.3 man-hours, ten times less than the about 3 man-hours 50 years ago. The cement industry was always able to produce cement at a competitive price, despite of much higher investment cost for the modern production unit.

The companies involved in cement production experienced significant changes in the past 100 years: from small, usually family owned companies with one single cement plant, to multinational large corporations with a great number of plants. This process was slow, gradual, and with somewhat faster development in the past 20 years during which the few market leaders developed their strong position worldwide. This process has changed the business culture of the industry entirely, and the types of activities and priorities have also changed. In the earlier days the technical issues were decisive in the small companies. Today the focus is on different, business oriented issues (profitability and market leadership), such us raw material resources management, logistic, maintenance, procurement, optimization and standardization of the production, marketing, information technology, automation, rationalization, and technology transfer within the company. All these issues are of great potential for cost savings and improved financial performance in general, making the large companies more competitive compared to small ones. In the United States, at present, perhaps over 80% of cement plants are owned by foreign multinational organizations.

We prefer to avoid discussion of the numerous negative aspects of large-scale companies. They exist – we can assure you! Just some key words to remember: lower flexibility, sluggishness, protection of the existing technologies (serious barrier to innovation), excessive centralized organization, and complex communications (excellent in theory, poor in practice).

In general, the progress accomplished by the cement industry during the past 100 years is magnificent, but it has been always very slow compared to other industries. It took – and still does take - many years or even decades until a new idea, invention, or methodology is successfully introduced into practice.

The Concrete Industry

During the past 100 years, both concrete science and technology experienced great developments as well. Today, concrete is the most widely used building material: we find concrete in all kinds of residential, industrial, agricultural and public buildings, in underground structures, all kind of transportation structures such as roads, pavements, railroads, airports, bridges and tunnels, in structures for water and sewage treatment, dams, offshore structures, etc. The total world production of concrete is estimated to be about 10 billion cubic meters, representing a value of about $600 billion! The business is highly fragmented into various segments, branches, and companies with different interests and often not-well defined responsibilities. This "monster" business shows the same signs of sluggishness as the cement industry, and there is even less coordination of activities and slower acceptance of new technical developments than in the cement business. The construction industry, especially the concrete part of it, is a typical conservative industry with steady but very slow evolutionary progress.

Irrespective of the problems, great achievements and progress were made in the last century. The main milestones which slowly but definitely changed and moved the industry to a leadership position are: introduction in 1913 of ready mix concrete, the "w/c law" introduced by D. Abrams in 1917, the lightweight concrete and shotcrete since 1920, use of pumps and vibrators in concrete placing since 1925, development of prestressed concrete by E. Freysinnet in 1930, invention of air entrained concrete for better freeze-thaw resistance in 1938, use of silica fume since 1952, expansion of precast concrete technology, use of chemical admixtures - in particular superplasticizers since 1955, slip forming of pavements and buildings since 1955, and use of fiber-reinforced concrete since ca 1960.

In recent years, the most important developments include the introduction of so-called *high performance concrete* in high-rise buildings, offshore structures, large-span bridges, and other sophisticated structures; however, as is always the case, even these new developments resulted in new challenges that need to be resolved [2]. The *self-compacting* or *self-levelling concrete*, allowing speed-up of the construction process, is another new technology with significant growth rate and promising future. Both these new technologies were made possible primarily due to the introduction of superplasticizers and microfillers (e.g., silica fume).

Another topic that became significant in recent years is recycling of demolished/used concrete. A number of research projects deal with this issue and the developing technology is becoming important from both the environmental and economic points of view.

From the point of view of concrete as a material, however, there were only few real break-through inventions realized in the past hundred years: air entrainment, high-range water reduction, fiber reinforcement, and introduction of microfillers.

Versatility: The reasons for the wide use of concrete are not difficult to define: the world wide availability of raw materials, relatively low price, the availability of a wide range of application technologies, and wide range of properties that can be achieved by clever use of the available technologies. Concrete is the most versatile material in terms of density, mechanical strength, thermal conductivity, etc. Its limited performance in tensile strength is the only real drawback when considering the versatility requirements. The extensive use of a great number of concrete types in various construction systems using a variety of production methods, illustrates the extraordinary versatility of concrete. None of the competitive building materials reaches such a wide use in man-made structures.

Science/Chemistry of Hydration and Degradation: The cement and concrete science contributed a great deal to the success story of cement and concrete industries. Meaningful progress has been made in better understanding of the hydration/hardening and degradation mechanism of concrete. New sophisticated tools, such as environmental SEM, NMR, and other advanced methods, were introduced to investigate the hydration processes and microstructure development of concrete. The valuable scientific work, reflected in numerous publications, international conferences, and in-house reports, deals mainly with new technologies/products and serious problems of concrete durability. The emphasis during the past 40 years was on the issues of alkali aggregate reaction, sulfate attack (lately especially on the heat-induced internal sulfate attack or delayed ettringite formation, DEF, and thaumasite formation), reinforcement corrosion, and freeze-thaw and de-icing salt resistance. More sophisticated physico-chemical research methodology has been adopted into the cement chemistry field and use of computers enabled development of predictive durability models. In the area of new technologies developed during the last 20 years, we must again mention the valuable scientific work on superplasticizers, fiber reinforcement, and rheology of concrete, all enabling the introduction of high performance and self compacting concrete into practice.

Concrete durability issues remain the most important one, as truly major investments are dedicated to maintenance, repair, and rehabilitation. Although only a very small fraction of concrete structures must be demolished or repaired as a result of catastrophic durability problems, it is still a significant loss to national economies and often hurts not only the owners but the concrete business as well.

3. DEFICIENCES AND BARRIERS

Notwithstanding the above-highlighted successes, we still face *problems* that are, in politically correct language, referred to as *challenges*. The technical and institutional problems we face today are not new; they were discussed already 50, 30 and 10 years ago … and even earlier. Let us quote from some of the conveniently forgotten literature:

G.M. Idorn, *Durability of Concrete Structures in Denmark*, Technical University of Denmark, 1967

> *"...durability of concrete ... can be considerably improved by only small* ***improvement of the initial quality*** *... suitable combinations of low w/c-ratio, careful compaction and correct curing conditions. Through these precautions we can obtain ... concrete which will eliminate ageing as a significant factor in deterioration ..."*

McLaughlin et al., *Concrete - Today and Tomorrow*, Civil Engineering, August 1969

> *"... qualitative demand ... will call for* ***radical revisions*** *in architectural concepts and structural design as well as in construction* ***methods and materials****."*

> *"As in the past, much of the* ***change will*** *be impelled by business and competitive considerations, and will* ***come from the users*** *of concrete."*

Klemm et al., *Cement Research: Boon or Boondoggle?*, Rock Products, February 1977

> *"...problem exists not only in the rules and restrictions imposed upon our industry, but in our own lack of understanding."*

> *"Intelligent and well-managed research and development are not antagonistic to increased production and profitability. They are and should be important parts of the daily life of a company, especially of a company looking ahead."*

National Research Council, *Concrete Durability: A Multibillion Opportunity*, National Materials Advisory Board, Washington, DC, 1987

> *"The false sense that concrete is an almost indestructible material results in an* ***undue emphasis on construction cost rather than life-cycle cost of structures****."*

> *"Not one chemistry or chemical engineering department at a U.S. university can be identified that is interested in the complex problems of concrete chemistry."*

> *"For financial, legal, and other reasons, training programs have been curtailed, contributing to* ***the decline in quality of the already "low-technology" concrete labor force****."*

> *"The technology of service-life prediction in concrete is advancing very slowly, even though the tools for generating and organizing knowledge (e.g., smart instruments, data base management systems, expert systems) are advancing rapidly."*

Robert E. Philleo: *Concrete Science and Reality,* in Materials Science of Concrete, Volume I, The American Ceramic Society, 1995, pp.1-8

> *"Most problems relevant to concrete science are those associated with high w/c, which belongs to the past, or those associated with low w/c, which belong to the future."*

Not much has changed since the above statements were published: we still occasionally violate some of the basic rules of concrete making.

Allow us to comment on some of the pertinent issues. We're fully aware that the challenges and their possible resolutions are complex, thus some of the forthcoming statements may be provocative oversimplifications ... but they are facts and we want to be provocative.

VECTOR DIAGRAM: TECHNOLOGY ACCEPTANCE IN THE CONCRETE INDUSTRY

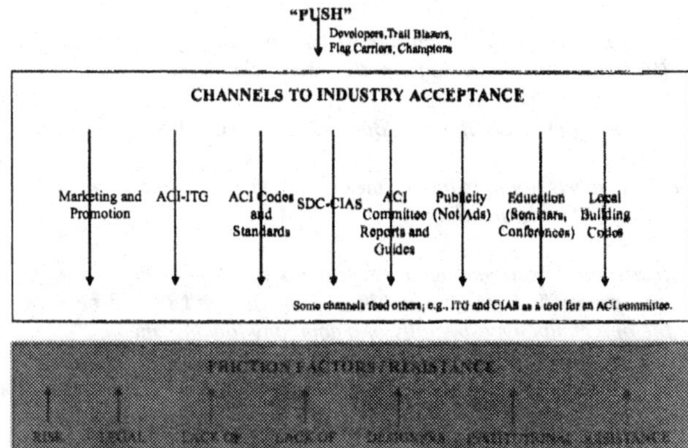

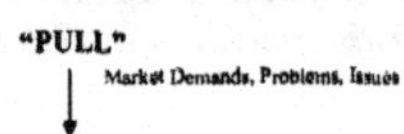

Educational problems

With the exception of a few well-known academic establishments, undergraduate and graduate education of civil engineers in concrete technology is marginal. Concrete as an engineering material is thus considered as a *virtual* material. It is implemented in the built environment without full consideration of its performance characteristics or material properties, often by those with little if any understanding of the basic physico-chemical principles upon which its mechanical behavior in a structure is based. The average civil engineer does not understand the fact that both development and deterioration of concrete properties, including permeability, modulus of elasticity, strength, and durability, is based on ***chemical*** processes, and consequently does not appreciate that this fact must be considered as fundamental during concrete design and subsequent processing. While civil engineers thus pretend to understand concrete chemistry, they are not alone in venturing into areas beyond their proper capacity. Geologists imply knowledge of structural issues; chemists assume the role of experts in life-cycle prediction. None of these groups is well skilled in materials science, but all of them, surprisingly, hold themselves out in litigation as experts on use of materials. And, of course, attorneys purport to hold an unqualified and unsurpassed omnipresence in all of the above fields.

This lack of knowledge thus translates into all sorts of institutional problems, including inadequate education of the work force, low quality of technical management in the industry (including R&D), inadequate respect for creative innovation, re-"invention" by academic researchers of known phenomena using somewhat advanced instrumentation and computational techniques, lesser overall quality of publications and research journals, a general decrease in expectations with respect to product quality, questionable expertise of "experts" in litigation, lowered industry reputation and reduced prestige of professional organizations and trade associations, and on, and on, and on.

Let us comment in more detail on one issue mentioned above, the issue of dissemination of data. It seems to us that the amount of published papers is inversely proportional to their quality. With a few notable exceptions, the submitted cement materials science manuscripts are not basic science papers, but repetitions of known facts – applied science papers at best! Using the latest version of an environmental scanning electron microscope (ESEM) or a nuclear magnetic resonance (NMR) apparatus to study yet another, thus far unknown fly ash, does not make the manuscript publishable. Every generation of scientists is guilty of repetitions and mistakes, but the recent explosion of inadequately researched and badly interpreted results goes too far, in our view. We need to find a better criterion for academic advancement than the presently-used "publish or perish" model – a model that confuses *quality* with the *number* of publications in "peer-reviewed" journals. Even this is violated sometimes, as we know of instances when peer-rejected papers were eventually published in the same or other journals. Peer review does not always guarantee quality! In other cases, scientifically incorrect manuscripts are accepted on the basis of past reputation of the authors. We need to tighten the belt – for writers, reviewers, and editors.

In our view, the outdated university tenure concept – originally introduced to protect non-technical academicians' freedom to promote politically and otherwise controversial views - is unnecessary and counterproductive. It leads to overall decrease in the quality of the academic establishment.

A related issue is the excessive number of international conferences and journals dealing with cement and concrete science and technology. We would like to mention just one of many examples. During the past 20 years, there were about 30 CANMET/ACI conferences on mineral components, superplasticizers, advances in concrete technology, and durability of concrete. All very topical and important issues (presented at most exciting and beautiful places around the globe) but … let's be honest (!), a much better coordination of conference activities, with lesser repetition and emphasis on "reinvention of the wheel," is badly desired[1]. The same is true for many other meetings and conferences, possibly including the present one!

The academic community contributed significantly to the progress and successes of the cement and concrete science and business. We should be proud of its contributions. However, one may criticize the excessive and often unproductive time and money spent on the explored issues, primarily caused by selection of less-than-critical topics and mismanagement of human and financial resources.

Coordination and Management of R&D

Cooperation between industrial and academic establishments with respect to innovative research and development is limited, and coordination of work between universities themselves, and even universities and governmental funding agencies, is lacking. Some of the failed "research consortia" are typical examples of miscommunication and lack of progressive leadership. More often than not, industry funds academic research, if at all, only for public relations or marketing reasons, and the projects given to academic consortia or individuals are only of marginal interest to the grantor. The really important R&D is done by the company itself, and this is justified by the proprietary nature of the desired product development and the obvious conflict with the academic "publish or perish" philosophy.

Only very few construction materials companies support innovative technology development. Most of the companies, in our view, do not want innovation as they are happy with the *status quo* and any new developments are perceived to be expensive and troublesome. Short-term vision is the rule. Others don't know what they want or need. Even in the most advanced construction materials-producing corporations, there is always a controversy about the need for innovation, even if it is otherwise accepted to be of high potential at the technical level. Why would a cement company undermine its economic existence by attempts to replace portland-clinker based cements? Why would an aggregate company promote use of recycled concrete? Some of the more innovating admixture companies are actually pro-research but, let's not forget, they often market very useful products that would not be needed if we would follow the best practices of concrete making.

Knowledge Transfer and Product Development

There is a tremendous reservoir of unused knowledge available to us. Why isn't it used? The reasons are multiple and complex, and are economic, technical, cultural, and political in nature. First of all, there is a serious language problem between academia and industry, between research and marketing people. As stated above, industry does not trust academia to do the application part of the product development, this resulting in incomplete communication of technical details. Moreover, the jargons of marketers and researchers are situated at two different planes, thus it is imperative to have forceful and technically skilled management to force the transfer of ideas between the two cultures. Most academic scientists do not appreciate the real

---

[1] *However, admittedly, thousands of young engineers and scientist working on behalf of this huge industry must go their own way, learn, publish, and make their own mistakes as we did. If an industry with an estimated turnover of $600 billions would spend only 0.2 % on R&D, then about $1.2 billion should become available.*

problems industry is facing and the industry is unsure what they need and can expect from the academia; it is like squeezing the square peg of marketing into the round hole of science.

Secondly, due to the inadequate appreciation of the latest technical advances, there is intellectual and economic resistance to new ideas. Use of construction materials in structures is a multidimensional issue and there are only few people in the engineering field and its management who can appreciate both the structural and chemical aspects of service life and its maintenance. Additionally, most of the decisions are based on economic considerations, not on technical merit. How can we then expect a ready mix producer to voluntarily comply with technical requirements, when he isn't required to have even the basic technical education that would allow him to understand the consequences of delivering an inadequate or misplaced product? How can a technically uneducated manager make technically creative decisions?

Also, there are new ideas and technologies available today that are often unappreciated and resisted by the older, more conservative generation that wasn't exposed to these advances. It happens in science and technology all the time: conservative authority inhibits progress! Critical new discoveries often come from younger people who are willing to break the established boundaries. Conservatism is a chronic disease afflicting academicians, structural designers, producers of materials, codes specifiers, field inspectors, and owners. *Status quo* is a simpler, less risky avenue. How many people in the field know the scientific significance of mixture proportioning and water-to-cement ratio (and of the resulting effects on transport phenomena) and truly understand their impact on engineering properties and service life of concrete? Can one expect innovation under these circumstances? As history records time and again, lack of understanding results in Philistine rejection of new technologies.

Thirdly, there are few if any penalties for misusing existing technologies, and no economic rewards for using technically and ecologically viable new ideas. Responsibilities are diluted, unknown, and avoided. The position of most professional organizations on this issue, including of ACI, is vague. No one wants to stand up to actually implement current knowledge or embrace true scientific innovation, lest one offend a customer, lose an important member or, worse yet, alienate corporate financial support. For example, the mission statement of ACI's Responsibility in Concrete Construction Committee states [5]: *"Make recommendations regarding the responsibilities of and the interactions among the principal parties involved with concrete design and construction"* and summarizes its activities as follows: *"Committee continues to survey engineering schools on hours spent teaching ethics/responsibility."* This seems to be an unacceptable and inadequate activity that skirts the issue and shirks the mission. The current ACI Guidelines discussing responsibilities in design and construction are vague and contradictory: the complex issue of responsibility is discussed in the spirit of avoiding rather than accepting responsibilities; the document does not discuss the ethical principles that are, or should be, of paramount concern. The committee should focus - as charged - on developing and recommending avenues for responsible protection of the industry's economic interests by implementing and promoting new ideas following the rules of engineering ethics and building codes while, at the same time, keeping in mind the well being of their primary customers, namely the <u>owners</u> of the constructed facilities. Measuring time spent in classrooms where standards that have not been established are supposedly taught shows only an unfortunate waste of the student's educational resources, not delineation and promotion of responsibility within the industry!

In our opinion, the formation by ACI of the Strategic Development Council (SDC), a forum for the most advanced ACI members to meet and promote new ideas, is an indirect admittance by ACI that a new or additional way of doing business is needed. The Council should be highly complimented for its work (incl. that on codes simplification and, hopefully, enforcement). ACI should continue to reorganize itself, to focus on engineering quality and education rather than becoming a trade/lobby organization defending their members against their own mistakes! Like in national politics, no bureaucracy voluntarily reorganizes itself! Therefore all of us must get involved and overcome these institutional barriers.

Tests and Standards

There are considerable barriers to innovation in the process of development of standards as we know them today. Standards often promote the *old* technology, and *new* standards, test methods, and technologies are torpedoed in committees and subcommittees composed of lobbyists (often less-rather-than-more knowledgeable) for one of many groups motivated by selfish private interest. Examples of this are the continuous fights in ASTM, ACI, and elsewhere for recognition of generally accepted methodologies and scientific techniques, including use of SEM as an important methodology for petrographic evaluation of concrete quality and adoption of the best analytical methodology for determination of sulfate levels in soil. Other examples abound.

As discussed by many, it is also our opinion that numerous existing test methods and standards are outdated and need to be reexamined and rewritten to incorporate the latest scientific advances **and** practical experience [1, 6-12]. At the same time, they have to be simplified ... not an easy challenge! Many present tests are based on laboratory experimentation under arbitrarily chosen environmental conditions, and thus are inadequate in describing the complex chemical and physical changes that lead both to development of concrete properties and also to their deterioration under the particular field conditions.

An example is the indiscriminate and uncritical use of compressive strength to characterize concrete quality, including its durability. Designers routinely use compressive strengths to assure structural integrity, without further consideration or specification to address long-term durability or service life. This unacceptable practice continues in spite of the fact that it has been well established and is known world wide that the compressive strength-to-durability relationship is flawed and unreliable. A structurally sound concrete is not necessarily durable. As an example, there is no simple or direct relationship between concrete matrix expansion caused by ettringite formation or alkali-silica gel swelling and strength. As a matter of fact, "durability" *per se* as a single performance characteristic is not a quantitatively definable property of concrete (or, for that matter, any other material). The Guide to Durable Concrete [13] defines durability as "*its ability to resist weathering action, chemical attack, abrasion, or any other process of deterioration. Durable concrete will retain its original form, quality, and serviceability when exposed to its environment.*" It is thus clear that "durability" involves chemical, microstructural, physical, and mechanical stability under various environmental conditions. Under such reality, testing a single parameter (such as expansion, compressive strength, or any other single parameter) and then using the results to describe the level of concrete "durability" under any or all conditions of use is simply wrong because the measurement is grossly inadequate by definition.

We would like to challenge ASTM and other standardization organizations that develop materials standards to reevaluate their *modus operandi* and to refocus their efforts on development of tests and standards that address real-world conditions and take into consideration the technical complexities of widely acknowledged deterioration mechanisms and environmental conditions. We need direct measurements of the basic properties of materials used in construction, not simply surrogate replacement measurements of vague and ambiguous relationships between, as an example, "durability" as compared to compressive strength. Specifically, there is need to develop reliable field tests for concrete permeability and diffusion coefficients, as most chemical (e.g., ASR, external sulfate attack, corrosion of reinforcement) and physical (e.g., freezing-thawing, erosion) mechanisms of concrete deterioration are to a large degree controlled by transport phenomena.

As well summarized a few years ago by S. Rostam [2], *"the growing complexity of structures and the demands for reliable performance and service life are continuously leading to more and more complex specifications ... this leads to increased cost of producing concrete* ...and *reduces the competitiveness of concrete ... a vicious circle ..." "Transparency and simplification in the requirements, increased standardization and acknowledgment of experience from practice when writing the specifications could improve competitiveness as well as improve long term quality of structures."*

Environmental challenges

With increasing population growth and economic globalization, the primary socio-economic pressures on the construction industry will be the environmental issues. These will strongly affect both the cement industry and the concrete industry. Just to list a few ongoing and upcoming challenges facing the cement industry:

- increasing cost of fossil fuels generates pressure to use less fuel, thus pressure to replace Portland cement clinker;
- generation of large volumes of carbon dioxide, oxides of nitrogen, and other pollutants, leads to pressure to cut clinker production and to increase use of waste products from other industries;
- pressure to eliminate hexavalent chromium in portland cement world-wide,
- need to maintain profit margins and product differentiation in spite the above challenges.

The concrete industry faces challenges too. Among others:

- need to use marginal, not well controlled concrete components such as recycled or marginal aggregate and less-rather-than-well characterized fly ashes, various industrial slags, limestone, etc.
- how to control the alkaline nature of concrete, esp. during its disposal;
- how to economically increase the (technical and environmental and ethical) quality of the work force?

The solution to these challenges won't be linear and will require close cooperation between the industry and governments at various levels, with active help from the academic establishment. New and possibly difficult solutions will be needed, and they will have to be accepted by the industry and the users in spite of the fact that many of the limitations we'll face will be perceived as limitations to our excessive "freedoms" with respect to "Mother Earth".

Regulations – Enforcement – Litigation

Most codes and standards *per se* are only minimum standards, thus enforcement of the best practices of design and construction is left to the contract between the construction participants: owners, developers, materials suppliers, architectural firms, and builders. Of course, these guidelines must be in compliance with local "culture," which may or may not always be memorialized in the form of a clear, certain, and binding contract.

It is unfortunate that the recent explosion of construction litigation - and the associated large-scale redistribution of U.S. dollars from insurance companies through law firms to the owners of damaged concrete structures - divides what should be a non-partisan scientific and technical community into antagonistic camps. Litigation is the result of cracks in our resolve to maintain and adhere to available knowledge of material qualities and best production practices ... attorneys have seen these rifts and have forced gaping chasms into our ranks.
While this chaos creates considerable noise, rising above the din of conflict we hear at least three truths ringing loud and clear:

a) owners of structures may have **legitimate reasons to complain** about the quality of construction and should have the opportunity to request reparations;
b) the construction community needs to **accept responsibility** in cases when mistakes were made rather that hide behind unreasonable and technically incorrect excuses; and
c) only those cases should be litigated where the alleged **damage is credible and is the result of clear violation of existing building codes and best practices of construction and design.**

While we do not object to the existence of two or more different but honest expert opinions, we are quite sadly disappointed that the "dirty" approaches common in political life have unfortunately infiltrated the concrete scientific and technical community, including the professional organizations. Professional colleagues avoid each other, publicly question each other's integrity at professional meetings, and sometimes claim opinions that even they cannot believe. Many times these represent unproved theories and "junk science" used only to bolster a weak argument.

Papers are published by self-centered partisan experts for the sole purpose of deflecting their own incompetence and discrediting opposing experts. By influencing editorial boards, publication of opposing views is sometimes prevented and opinions of only one side are accepted or even solicited. Committees of professional organizations are misused in an attempt to make changes in guidelines and standards that are hoped to benefit one side or the other in pending litigation. Battles are being fought for membership and chairmanship of committees and sub-committees that are perceived to be of indirect importance in litigation. In litigation, the existence or non-existence of a particular ASTM standard (independent of its quality and actual use in engineering practice) is often more important than the technical facts, and valid scientific methods are ignored because they purportedly are "*not accepted by the scientific community.*" ASTM is argued by some to be the sole representative and only benchmark of the "scientific community" – an entirely erroneous myth! Truth is being massaged, and scientific fact (indeed, science itself) becomes fuzzy. Clintonesque questions à la *"What is the meaning of 'is'?"* are being asked. Our mutual intelligence is being questioned. Money corrupts.

In a more cynical mode, one could ask: *Why shouldn't the construction defect attorneys use the stupidity and mistakes of the construction industry against it?* Legislators in most countries, including in the U.S.A., do the same thing against all of us routinely!

Professional organizations – if they really want to retain their professionalism and regain credibility they justifiably seek – should take the lead in convincing themselves that a cultural change is desired, if not to wholeheartedly pick up and carry forward the standard of proper science then at least to choose the right banner to follow. Many years ago, George Leyh, executive director of ACI, said at a meeting at NIST that "*We're not leaders, we're followers.*" ... and that's fine, so long as we follow the technically, ethically, and otherwise correct path!

4. CONCLUDING REMARKS

All of us should have a dream about the future; in this particular case, the dream should be about a prosperous, sophisticated, ethical, and economically and environmentally sound cement and concrete industry. The dream should be an ambitious one because the future, after all, does not belong to us but is ours to shape for the next generations, including our children and grandchildren. To succeed, we need to:

- Learn from the past and adapt positive approach to new knowledge; use this knowledge to develop new, environmentally and economically sound building systems (hurricane and floods resistant housing, etc.);
- Improve education at all levels by, among others, introducing a win-win dialog between academia (new knowledge) & industry (profitability) & professional organizations (societal impact);
- Concentrate on transfer of existing knowledge via hands-on demonstrations and mandatory rotation of people between academic research and industry;
- Develop additional, new approaches to standardization; introduce enforcement procedures that are protecting both the industry and the customers;
- Keep in mind the need for increasing product quality (materials, structures, systems), fight mediocrity and complacency, and clarify specific responsibilities of parties involved in the construction business.

Acknowledgments and apologies

We would like to express our thanks to numerous colleagues who discussed with us the ideas and examples discussed in the text. If some of our criticism hit too close to home, we would like to assure you that the similarities are purely coincidental. We would also like to apologize to those who feel badly because they were not included in our criticism.

References

[1] A. Bentur, *Cementitious Materials – Nine Millennia and a New Century: Past, Presence and Future,* Journal of Materials in Civil Engineering, 2002, pp.1-21

[2] S. Rostam, *Does High-Performance Concrete Provide High Performance Concrete Structures?,* presentation at the FIB International Symposium on High Performance Concrete, Orlando, FL, September 2000

[3] Portland Cement Association, *Innovations in Portland Cement Manufacturing,* 2004, 1354 pp.

[4] I. Odler, *Special Inorganic Cements,* E & FN SPON, London and New York, 2000, 395 pp.

[5] ACI Committees 2002-2003, American Concrete Institute, 2000

[6] P.K. Mehta (Ed.), *Cement Standards – Evolution and Trends,* ASTM STP 663, ASTM, 1978

[7] E. Farkas, *Concrete in the Year 2000,* 10th Conference on Our World in Concrete & Structures, Singapore, August 1985

[8] L. Spellman and J. Skalny, *ASTM Standards and Materials Research,* in Materials Science of Concrete, Volume III, The American Ceramic Society, 1995, pp.391-405

[9] R.D. Hooton, *Are Sulfate Resistance Standards Adequate?,* in Materials Science of Concrete: Sulfate Attack Mechanisms, The American Ceramic Society, 1998, pp.357-366

[10] J. Gebauer, *Notes on R&D in Cement and Concrete,* in Materials Science of Concrete: Cement and Concrete - Trends and Challenges, The American Ceramic Society, 2002, pp.149-154

[11] F. J. Young, *Bringing Concrete into 21st Century,* in Advances in Cement and Concrete, 2003, pp.1-8

[12] J. Skalny and G.M.Idorn, *Thoughts on Concrete Durability,* in Concrete Science and Engineering: A Tribute to Arnon Bentur, Proceedings PRO 36, RILEM Publications, 2004, pp. 223-229

[13] ACI 201.2R-01, *Guide to Durable Concrete,* ACI International, 2001

U.S. CONSTRUCTION AND CEMENT OUTLOOK

Ed Sullivan
Portland Cement Association
Skokie, IL, USA

1. Overview

The Portland Cement Association (PCA) has adjusted downward its near-term outlook for the U.S. economy. Real GDP for 2004 is now projected at 3.9%, compared to 4.4% previously forecast, and 3.4% for 2005 compared to 3.8% growth previously forecast.

The downward adjustment to the current forecast primarily reflects significantly higher oil price assumptions. PCA fully incorporates the likelihood of continued oil supply disruptions in the context of strong global demand conditions, resulting in a downward rigidity in current oil price levels. The higher oil price scenario will weaken overall economic growth. With higher oil prices, consumer spending will be partially compromised, inflation will run stronger, job gains will be smaller, and sentiment in both the consumer and business areas will be more sedate. Combined, these factors lead to roughly a 50 basis point reduction in PCA's previous forecast for real GDP growth.

The higher oil price scenario translates into three key adjustments to PCA's construction forecast. First, slower overall economic growth implies a more gradual recovery in capacity utilization and vacancy rates, and generally lowers the expected return on investment for most commercial properties. According to PCA's current scenario, the recovery in nonresidential construction activity is softened and delayed slightly. The summer forecast projected a 2.0% gain in 2004 nonresidential construction spending, followed by a 12.3% gain in 2005. The current forecast now shows a 0.1% gain for 2004 nonresidential activity, followed by a 9.9% gain for 2005.

Second, slower overall economic activity implies slower growth in employment. The burst in employment gains that materialized at the time of PCA's summer forecast has subsided dramatically. In the near term, net job gains are expected to range between 150,000 to 175,000 per month – resulting in 1.9 million net new jobs for 2004 and 2.1 million for 2005. These lower job gain estimates reflect roughly a 300,000 to 400,000 job growth reduction compared to our summer forecast estimates.

Slower job growth implies a more gradual improvement in states' tax base resulting in slower improvement in state revenue growth compared to PCA's summer projections. The scars from state fiscal problems, therefore, will fade somewhat more slowly than previously anticipated. With a softer recovery in state fiscal conditions, public construction activity is expected to be slightly more constrained. PCA now expects public construction will record a 1.5% reduction for 2004, followed by a 4.0% increase for 2005. This compares against a 1.4% 2004 and 4.3% 2005 gain contained in the summer forecast.

Beyond 2005, PCA has incorporated a higher TEA funding level. PCA's summer TEA funding assumption reflected an average of the Bush, Senate and House proposals – translating into a $280 billion bill beginning in fiscal 2005. Since the summer forecast, PCA's Washington office has concluded that the Bush Administration is willing to support a higher level of funding commitment to the new TEA – raising the funding level to $299 billion – a 6.8% increase over previous PCA assumptions. The higher TEA assumption has been integrated into the new forecast along with a fiscal 2006 starting point.

Finally, the higher oil price scenario tends to change the expected returns within financial portfolios and, as a result, the amount of funds that are committed to the bond market. Typically, economic recoveries are associated with rapidly improving profits, price/earnings ratios, and confidence in the equity markets. In this context, funds typically flow out of the bond market and into equities – diminishing the funds invested into the bond market and thereby acting as a factor raising interest rates in the bond market which is then reflected in mortgage rates.

Higher oil prices, however, not only increases investor uncertainty, but tends to depress the expectation of corporate profitability and hence price/earnings ratios within the stock market. In this environment, the typical outflow of funds from the bond market does not fully materialize – diminishing the rising of interest rates in the bond market and keeping the lid on increases in mortgage rates.

This implies that mortgage rates will not increase as quickly as projected in the summer forecast. PCA had expected that 30 year conventional mortgage rates would reach and exceed 6.5% at the beginning of the fourth quarter 2004. Keep in mind, the 6.5% mortgage rate is considered the "tripping rate" or the rate that will exert enough pressure on home affordability to result in significant declines in single family construction activity. In the current forecast, the "tripping rate" is not expected to materialize until the end of first quarter 2005 – thereby adding legs to the already strong single family construction run.

Aside from oil price impacts on the economy and the level and composition of construction activity, PCA has incorporated an upward adjustment in cement intensities for most nonresidential and some public construction sectors. Cement intensities measure the amount of cement used per level of construction spending. The increase in cement intensities is based on an improvement in the competitive conditions of concrete relative to steel. The rapid run-up in steel prices has not been matched by concrete, thereby improving concrete's relative competitive condition versus steel.

2. Oil Outlook

The outlook for oil prices continues to be marred by global supply disruptions. Iraq's ability to supply oil to the international markets has been dramatically hindered by ongoing acts of terror – halving its exports. Russian oil supply to international markets has been hindered by the loan default of Russian oil giant Yukos. Nigeria's supply to international markets has been hindered by labor strikes. Finally, the recent hurricanes have cut production dramatically in the Gulf.

Significant increases in demand arising from the global economic recovery and robust Chinese demand also characterize the market. These factors, coupled with speculators' actions, have pushed oil prices to more than $54 per barrel. Aside from speculators' actions, the factors that gave rise to the increases in oil prices are likely to remain in place for the near future. As a result, PCA has adjusted upward its assumptions regarding the price of oil. Through 2005, oil prices are expected to stay near the $45 to $50 range. Thereafter, a gradual decline to the $30-$35 per barrel range by 2007 is anticipated. It should be noted that the longer term price assumption is in the upper end of OPEC's new target range.

Top Oil Producers
- Million Barrels Per Day

1	Saudi Arabia	10.1
2	Russia	9.1
3	United States	8.7
4	Iran	4.0
5	Mexico	3.9
6	China	3.6
7	Norway	3.3
8	Canada	3.2
9	Venezuela	2.8
10	United Arab Emirates	2.8

Average Daily Production 1st Half 2004

The higher oil price scenario will weaken overall economic growth. In the context of higher oil prices, consumer spending will be partially compromised, inflation will run stronger, job gains will be smaller, and sentiment in both the consumer and business areas will be more sedate.

PCA's latest U.S. economic outlook contains each of these adjustments. The new oil price scenario cuts overall real GDP growth by roughly 50 basis points in 2004 and 2005, pushing real GDP projections to 3.9% in 2004 and 3.4% in 2005, compared to 4.4% and 3.8% previously forecast. It is important to point out that the current high oil price scenario is placed in the context of a relatively healthy overall economy. The oil price increases, therefore, will not lead to a recession – but rather a moderation of relatively strong economic growth fundamentals.

Oil Price Impacts

	Summer Projections 2004	2005	Fall Projections 2004	2005
Oil Price Per Barrel Avg	$37.17	$31.23	$42.15	$48.78
Real GDP Growth	4.4%	3.8%	3.9%	3.4%
Net Job Creation, Million	2.3	2.5	1.9	2.1
Consumer Sentiment	105.9	115.0	98.2	111.5
Consumer Spending Growth	3.8%	3.6%	3.4%	3.2%
Nonresidential Construction Growth	76.0	79.8	75.9	79.4

Consumer Spending: The underlying fundamentals of consumer demand remain relatively strong. While consumer affordability and sentiment have recently declined, they remain strong from a historical perspective. Consumer balance sheets continue to improve as evidenced by declines in delinquency and personal bankruptcy rates. Job creation and access to low rate home equity markets will continue to provide fuel for continued consumer spending growth.

Some economists have suggested that the recent slowdown in second quarter consumer spending is a precursor to significantly more sedate overall economic growth for 2004 and beyond. PCA does not share this viewpoint. First, the economy rarely grows at an even pace throughout all sectors of the economy. Little weight should be granted to uneven monthly variations from longer term cyclical trends. Second, transition periods are typically characterized by pauses in growth. Consumers' source of funding is now undergoing a transition process reflected in the dramatic rundown in home refinancing and will eventually be replaced by stronger income and job gains. In our view, the recent slowdown in retail sales activity reflects a pause associated with this transition – not a sustained trend.

While the fundamentals of consumer spending remain on solid ground, PCA's higher oil price scenario forces a moderation in consumer spending growth. Higher gasoline and home heating bills will force consumers to reallocate funds away from consumer spending and toward energy needs. Depending upon the severity of the upcoming winter home heating season, the shift could be considerable. Compared to PCA's summer forecast of a 3.5% to 3.8% consumer spending growth rate anticipated during the winter season, a 3.2% to 3.4% growth rate is now anticipated. The combination of warmer weather and some erosion in energy prices suggests a resumption of accelerated consumer spending activity during the second quarter of 2005 and beyond.

Investment Spending: Improvement in the expected risk adjusted ROI, increased availability of funds for investment, and the release of pent-up investment demand form the basis of PCA's optimism regarding the investment sector.

The acceleration of investment spending growth during the remainder of 2004 is predicated on the sustained pattern of relatively strong GDP growth in past quarters that should reinforce business decisions to invest, providing additional support to already strong foundations for strong investment spending growth.

The higher oil price scenario, however, will adversely impact investment spending activity. The marginal slowdown in overall economic activity will slightly reduce expected return on investment calculations. In addition, business sentiment will be adversely impacted. Combined, these factors will take some of the edge off an otherwise strong investment environment. Nevertheless, PCA still expects investment spending will record double digit growth gains during 2004 and strong growth in 2005.

Oil and the Prospects of Recession: Some analysts note the relationship between the run-up in oil prices during the 1970's and 1980's and resulting recessions. These analysts conclude that the recent run-up in oil prices have fated the U.S. economy to a recession in 2005. Higher oil prices will slowdown overall economic growth. If large oil price increases are placed in the context of an already weakening economy, then a recession is possible – but only in the context of an already weak economy. This is not the context in which today's high oil prices are set. Today's economy rests upon relatively strong foundations such as low interest rates, low inflation rates, and an improving labor market. Higher oil prices will act only to shave 50 basis points of anticipated general economic growth rates and are not expected to result in recession conditions.

3. Construction Outlook

PCA expects U.S. inflation adjusted construction spending to increase by 3.7% in 2004 to $724 billion ($1996). For 2005, construction spending is expected to reach an inflation adjusted level of $745 billion or 2.9% growth. These growth rates are made more impressive when one considers that the construction market has been performing near historical peaks.

The composition of construction activity during the 2005-2008 period will be much different than the composition of construction activity during the 2001-2004 period. Until this year, the economy had been characterized by weak economic conditions, staggering job losses, and extremely low interest rates. These conditions led to dramatic declines in capacity utilization, an increase in vacancy rates, reduced state revenue collections, mammoth state deficits, and extremely favorable mortgage rates. Favorable mortgage rates resulted in residential construction as the growth leader in construction – led by single family builds. Economic weakness and the corresponding run-up in industrial and office vacancy rates resulted in enormous declines in nonresidential activity.

Changing Composition of Construction Spending Growth

2001-2003

Low Interest Rates, Weak Economy

- **Growth Leader: Residential**
  - Low Interest Rates
- **Public**
  - State Tax Revenues Hurt by Anemic Economic Growth
- **Growth Laggard: Nonresidential**
  - Weak Economy

2004-2007

Rising Interest Rates, Strong Economy

- **Growth Leader : Nonresidential**
  - **Strong Economy**
- **Public**
  - **State Tax Revenues Recovery Due to Strong Economic Growth**
- **Growth Laggard : Residential**
  - **Rising Interest Rates**

Finally, deficit issues prompted state governments to postpone construction projects. During the 2001-2003 period, residential was the growth leader with nonresidential and public spending growth laggards.

In retrospect, 2004 represented a year of transition for the U.S. construction market. A year that state deficits, utilization rates and vacancy rates stabilized and began the process of healing – setting the stage for recovery in public and nonresidential recoveries next year. The year also served to set the conditions for rising mortgage rates to materialize in 2005.

As the economy embraces sustained, relatively strong, economic growth during the 2005-2008 period, interest rates will rise – prompting an eventual retrenchment in single family construction. At the same time, vacancy rates within the industrial and office sectors will gradually retreat – providing the basis of recovery for the nonresidential sector. Finally, recent tax increases levied by many states during time of financial distress will meet a recovery in job creation – generating a rather rapid recovery in state revenue collections, an erasure of state deficits, and eventually a sustained recovery in public construction spending. The improvement in states' fiscal picture will coincide with a renewal of TEA at higher levels. During the 2004-2008 period, nonresidential and public spending is expected to assume the mantel of growth leadership and residential activity will become the growth laggard (although maintaining a strong level when measured against history).

4. Single Family Construction & Mortgage Rate Outlook

Through the first nine months of 2004, single family building starts increased 9.8% over record 2003 levels. While PCA projected strong single family construction activity would prevail during 2004, we failed to forecast such large increases over 2003 levels that have transpired year-to-date. PCA expects single family construction activity will remain quite strong until conventional 30-year mortgage rates top 6.5%.

In the summer forecast, PCA expected that 30-year conventional mortgage rates would reach and exceed 6.5% at the beginning of fourth quarter 2004. According to PCA research, the 6.5% mortgage rate is considered the "tripping rate" or the rate that will exert enough pressure on home affordability to result in significant declines in single family construction activity. In the current forecast, the "tripping rate" is not expected to materialize until the end of the first quarter 2005 – thereby adding legs to the strong single family construction run.

Part of the reason mortgage rates have not increased to the levels expected during 2004 is a result of low inflation rates, easing demand in refinance activity, and the relative performance of the bond market versus the stock market. Typically, for example, economic recoveries are associated with rapidly improving profits, price/earnings ratios and confidence in the equity markets. In this context, investment funds typically flow out of the bond market and into equities – diminishing the funds invested into the bond market and thereby acting as a factor raising interest rates in the bond market - reflected in higher mortgage rates.

In the context of higher oil prices, investor uncertainty has been increased and the expectation of corporate profitability and hence price/earnings ratios within the stock market have been depressed. In this environment, the typical outflow of funds from the bond market does not fully materialize – diminishing the rising of interest rates in the bond market and keeping the lid on increases in mortgage rates.

PCA expects only a gradual erosion of the factors that have kept mortgage rates from rising at the rates previously anticipated. PCA now expects the "tripping rate" of 6.5% will not be reached until the second quarter of 2005. Until that time, the single family construction sector will perform at rates significantly higher than previously projected. In the current forecast, PCA has increased 2004 single family starts to 1.6 million (3% upward adjustment) and 2005 starts to 1.489 million (5.1% upward adjustment). This adjustment adds roughly 1.3 million tons to 2004 and 2005 cement demand.

5. Multi-Family Construction Outlook

The multi-family sector of the residential market is composed of two key segments, namely the rental sector (accounting for roughly 80% of total multi-family construction) and the condominium sector (accounting for the remainder). The outlook for each sector is completely different. The rental market, for example, typically thrives during a high interest rate environment and is exactly the opposite for the condominium market.

The outlook for multi-family rental construction activity has been adversely impacted by the low mortgage rate environment. Low interest rates have decreased the spread between the average monthly mortgage payment and average monthly rent - enabling more apartment dwellers to move into the single family sector. As a direct consequence of this phenomenon, apartment vacancy rates rose and now stand at 10%. Vacancy rates will not begin to decline until the draw from the single family market subsides. The process of multi-family rental demand recovery is not expected to begin until mortgage rates exceed 6.5%. A significant improvement in multi-family rental investment is not expected to materialize until vacancy rates are reduced to at least 8%. Such a threshold is not expected to be reached until 2006. Even when these conditions are set in place, the revival in rental multi-family construction is expected to be very modest. In specific regional markets very favorable demographic conditions will offset the adverse conditions that plague the national multi-family rental outlook.

The condominium market will generally parallel the directional movements of the single family market. There are, however, qualifications. Condominium ownership has a higher skew to the retirement age population. Retirees' income is typically tied to interest performing assets. While the rise in interest rates will lead to an increase in condominium monthly payments, it will simultaneously increase retirees' income – diminishing the erosion in affordability and hence the decline in condominium demands. Nevertheless, rising interest rates will force a decline in condominium demand in most regional markets – even in the face of favorable demographics.

6. Nonresidential Construction Outlook

Through the first nine months of 2004, nonresidential construction activity has averaged a seasonally adjusted annual rate of $121 billion – roughly in-line with PCA's spring and summer projections. Only minor adjustments, therefore, have been incorporated into PCA's forecasts for 2004. In the context of higher oil prices, however, slower than previously expected overall economic growth will materialize during 2005. This implies a more gradual recovery in capacity utilization and vacancy rates and generally lowers the expected return on investment for most commercial properties. According to PCA's current scenario, the recovery in nonresidential construction activity is softened and delayed slightly. The summer forecast projected a 2.0% gain in 2004 nonresidential construction activity, followed by a 12.3% gain in 2005. The current forecast now shows a 0.1% gain for 2004 nonresidential activity, followed by a 9.9% gain for 2005.

7. Public Construction Outlook

Through the first nine months of 2004, public construction has averaged an inflation adjusted $170 billion seasonally adjusted annual rate. The year-to-date performance is roughly in-line with PCA's summer forecast estimates and therefore requires only minor adjustments for the current forecast for 2004.

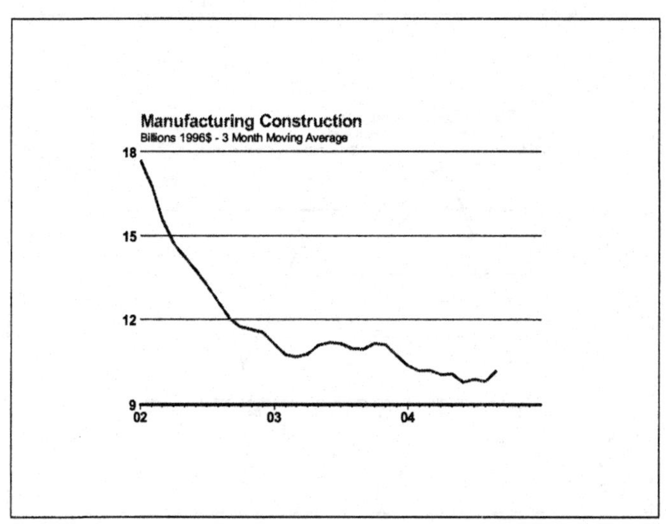
Manufacturing Construction
Billions 1996$ - 3 Month Moving Average
18
15
12
9
02
03
04

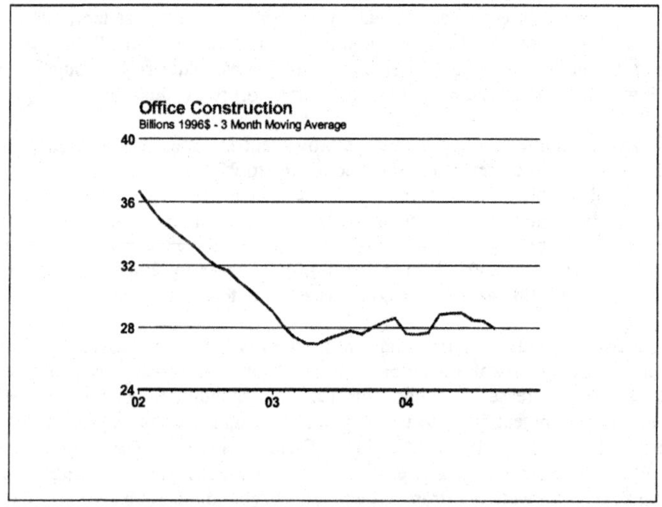
Office Construction
Billions 1996$ - 3 Month Moving Average
40
36
32
28
24
02
03
04

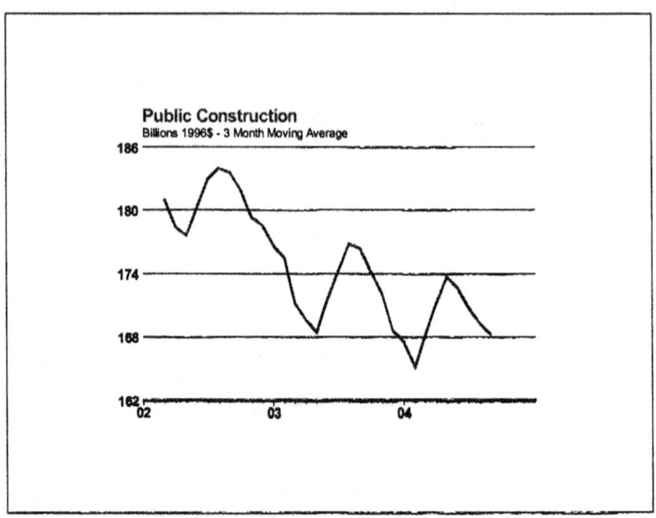

Beyond 2004, however, slower overall economic activity due to the higher oil price assumptions implies slower growth in employment. The burst in employment gains that materialized at the time of PCA's summer forecast has subsided dramatically. In the near term, net job gains are expected to range between 150,000 to 175,000 per month – resulting in 1.9 million net new jobs for 2004 and 2.1 million for 2005. These lower job gain estimates reflect roughly a 300,000 to 400,000 job growth reduction compared to our summer forecast estimates.

Slower job growth implies a more gradual improvement in states' tax base and, as a result, a slower improvement in state revenue growth compared to PCA's summer projections. The scars from state fiscal problems, therefore, will fade somewhat more slowly than previously anticipated. In the context of a softer recovery in state fiscal conditions, public construction activity is expected to be slightly more constrained. PCA now expects public construction will record a 1.5% reduction for 2004, followed by a 4.0% increase for 2005. This compares against a 1.4% 2004 and 4.3% 2005 gain contained in the summer forecast.

By 2006 the fiscal conditions surrounding most states will have improved dramatically. The improvement in states' fiscal condition will result in an increase in public construction expenditures. The increase in public construction expenditures reflects not only a more favorable fiscal environment, but also the release of pent-up construction projects that have been put to the side during an environment of harsh fiscal constraints. The combination of states' fiscal budget improvements and the release of pent-up public demand is expected to push public construction activity to relatively strong growth rates beginning in 2006 and carry throughout the remainder of the forecast horizon.

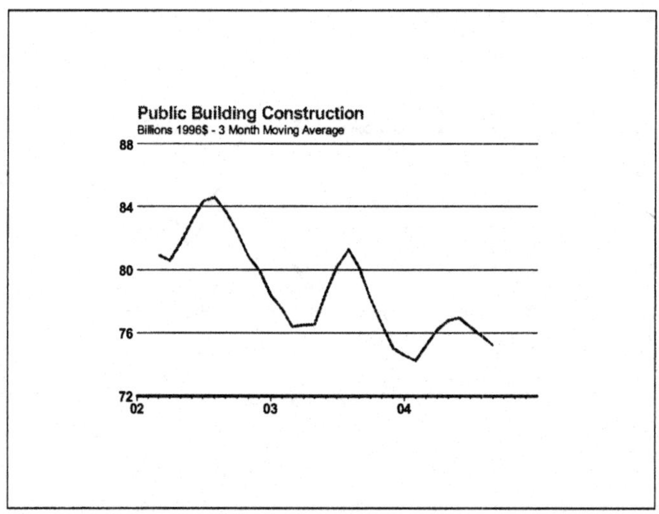

PCA has also incorporated a higher TEA funding level into the fall forecast. PCA's summer TEA funding assumption reflected an average of the Bush, Senate and House proposals – translating into a $280 billion bill beginning in fiscal 2005. Since the summer forecast, PCA's Washington office has concluded that the Bush Administration is willing to support a higher level of funding commitment to the new TEA – raising the funding level to $299 billion – a 6.8% increase over previous assumptions. The higher TEA assumption has been integrated into the new forecast along with a fiscal 2006 starting point.

8. Cement Intensities Outlook

The favorable outlook for construction activity is expected to coincide with an ongoing increase in cement intensities – adding to overall cement demand strength. Aside from PCA promotional efforts, several other factors will shape the level of cement intensities in the near term.

First, business cycles impact nonresidential cement intensities. During cyclical downturns, investors tend to be less committed to building large nonresidential projects – thereby reducing the cement intensity per billion dollars of construction. Investor retraction of the larger projects is tied to lower expected ROI's. During this period, smaller projects, with lower cement intensities, tend to account for a larger share of the nonresidential construction arena. As the economy recovers, the higher expected ROI's materialize yielding a return to larger scale projects and a recovery in cement intensities.

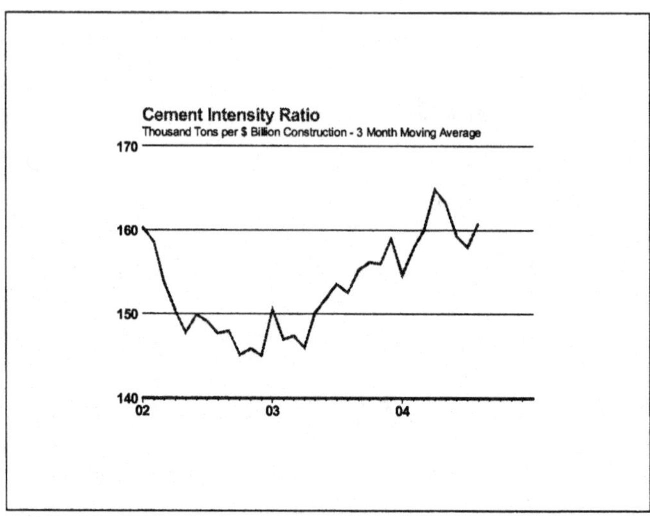

Second, the price competitiveness of concrete has dramatically improved against steel building products. Steel mill prices have increased 43.3% during the past year. The increase in steel mill prices is tied to scrap shortages largely induced by a dramatic increase in demand from China as the country undergoes a massive construction effort in preparation for the 2008 Olympic games. Given the nature of the demand increase, PCA believes that the recent run-up in steel prices will not be reversed anytime soon.

As a result of the run-up in steel prices, the relative price of concrete products to steel mill products has been cut by roughly 50% of the relative price that existed in 2002. Given enough time lag, this will lead to a substitution of cement for steel in some construction projects. The substitution is expected to be focused among smaller sized projects with a bit more design flexibility and a higher sensitivity to cost changes. This phenomenon will encompass not only nonresidential activity but public sector intensities as well.

According to PCA estimates, if the current relative price relationship between concrete and steel is sustained for a full year, nonresidential and public cement intensities will improve adding as much as 1.5 million tons to nonresidential cement demand and another 600,000 tons to public cement demand. PCA has incorporated these increases into our current forecast.

Third, residential cement intensities are also expected to rise due to sustained increases in the size of home as well as code changes in key states. Measured on a square foot basis, single family homes have been growing roughly 1% to 2% per year. This implies slightly higher cement requirements per single family home and thereby an increase in cement intensity per home. Building code changes, particularly in California and hurricane hit Florida, are also expected to act as an accelerant to residential cement intensities.

These positive impacts on cement intensity are weighed against a serious adverse impact on cement intensity. PCA expects highway cement intensities will ease throughout the forecast horizon compared to 2002-2003 levels. According to the apparent use revisions, highway intensities have increased 13.5% since 1999 and by 7.8% last year alone. Highway cement intensities stand at a 20-year high. PCA believes that the increase in highway intensities, particularly in recent years, is explained by the composition of highway construction. Given the budget distress faced by state governments, many local road projects were canceled. These projects tend to have a high asphalt intensity and low cement intensity. Federally aided highway projects have recorded only small declines. These programs tend to have a much higher cement intensity. As a result of the mix changes within overall highway expenditures, the cement intensity has increased. Going forward, however, will largely reverse many of these increases in highway cement intensities. As the budgetary problems facing state governments fade, more and more local road projects with the lower cement intensities will arise. A small downward push on overall highway cement intensities has been integrated into the PCA forecast.

9. Cement Shortage Assessment

Tight cement supply conditions now prevail in portions of 35 states. Since PCA's last survey, Texas, Arkansas, Missouri, Indiana, Ohio, Pennsylvania, New Jersey, Delaware, Vermont, New Hampshire, Arizona, New Mexico and Utah have been added. Nebraska has been removed from the list. It should be kept in mind that not all portions of each state are characterized by tight supplies. Northern Ohio and Indiana, for example, reportedly have ample supplies. Tight supplies in the southern portion of these states result in PCA's map showing tight supplies. The point here is the methodology used in the PCA shortage map tends to exaggerate the national shortage assessment.

Where cement is in short supply, the reasons are typically twofold: strong cement demand has materialized due largely to strong residential construction activity, and not enough ships are available to bring in imported cement.

In PCA's summer forecast assessment, a scenario was presented where an easement in shortage conditions could potentially be achieved in the fourth quarter. A critical ingredient of that assessment centered on an easement in fourth quarter demand brought about by rising mortgage rates. In the current forecast, PCA has lowered the projected near-term path of mortgage rates. This suggests that the vibrant single family housing sector will show more strength than previously expected – at least through the first quarter of 2005. As a result, a significant fourth quarter easement in shortage conditions brought about by demand is unlikely.

With PCA's current mortgage rate forecast, single family construction activity is now expected to ease in the second half of 2005. It is wrong to assume that this just implies a re-timing of potential easement in shortage conditions to the second half of 2005. By the second half of 2005, PCA expects relatively strong growth in nonresidential construction activity will emerge, cushioning the moderation in cement demand. PCA's foregoing demand assessment points to a continuation of relatively robust conditions – even in the context of higher oil prices and more moderate overall economic growth conditions.

Supply conditions hold the key to determining growth in the United States cement market for 2004 and beyond. Domestic production has been stretched to its limits and inventory levels have been squeezed tight. Further increases in supply, achieved from domestic operations, is limited. According to our estimates which incorporate capacity expansions, operating rate, and inventory assumptions, domestic supply is expected to increase by 2.3% in 2004, followed by 1.8% and 1.7% in 2005 and 2006 respectively. Keep in mind, these estimates reflect operating rates in excess of 98% and, therefore, may contain modest downside risks.

A key factor to achieve enough supply to meet ongoing, strong demand conditions centers on the level of cement imports that can be brought to the United States' market. Import projections are, in turn, are dictated by assumptions regarding international demand for ships and foreign availability of cement for export to the United States. While PCA takes an informed position with this regard, forecasts regarding United States cement imports continue to hold a high degree of risk. Furthermore, since PCA's forecast can be viewed as a supply constrained projection, uncertainty regarding imports impact overall United States cement consumption growth estimates.

For most of 2004, import volume has been constrained by the availability of dry bulk carriers and the availability of foreign supplies of cement. PCA uses freight rates as an indicator of conditions that characterize ship availability. During the fourth quarter of 2003 through February 2004, shipping rates increased more than twofold. The run-up in freight rates reflected extremely tight ship availability conditions. These tight conditions were brought on largely by the demand posed by rapid Chinese economic growth as well as a worldwide economic recovery

Capacity Expansion Estimates
(Thousand Metric Tons)

Clinker Capacity Expansion

	Start	Net Change	End	Growth
2002	-----	-----	91,146	
2003	91,146	750	91,896	0.8%
2004	91,896	1,925	93,821	2.1%
2005	93,821	525	94,346	0.6%
2006	94,346	1,500	95,846	1.6%
2007	95,846	4,100	99,946	4.3%
2008	99,946	4,850	104,796	4.9%

During the late spring and early summer, freight rates dropped from $49 per ton to $26 per ton. This signaled an easement in ship availability conditions. The improvement in freight rates and ship availability is attributed to Chinese economic policies aimed at reducing runaway demand conditions. The decline in freight rates implies that improved ship availability materialized around this time. Under normal times, it takes two months to set up a foreign order for cement and have it arrive at a U.S. port. Recent stress in shipping markets has extended that time to as much as four months. Given the time lag to bring in foreign cement, PCA did not expect imports to show meaningfully higher levels until September and estimated import volumes to run at a 27 million ton SAAR during the fourth quarter - bringing year total import volume near 25.5 million tons.

At the time, many considered the PCA outlook regarding import volumes as optimistic. Recent data suggests that PCA may not have been optimistic enough. August import volumes recorded a 31.0 million ton seasonally adjusted annual rate – the highest level in three and one-half years and the second highest August in history (August 1999 was slightly higher). This one data point raises the question whether August's performance can be sustained or is it an outlying data point?

PCA believes that import volumes will increase and probably exceed the summer estimates of a 27 million ton SAAR for the fourth quarter of 2004 and first quarter 2005 and has increased its 2004 and 2005 projections. Nevertheless, the 31 million unit SAAR recorded in August, is viewed as an outlier and is considered unsustainable.

For the past several months, freight rates have increased and now stand at $33 per ton. This implies that a re-tightening in shipping conditions has materialized and suggests a moderation in the improvement in cement import volumes going forward.

PCA expects imports will reach 26.2 million tons for 2004 or a 12.4% increase over 2003 levels. This reflects year-to-date information and an assumption of a 27.5 million ton seasonally adjusted annual rate for the remaining four months of 2004. For 2005, PCA has ratcheted up its import projections to 28.5 million tons, or an 8.8% increase over projected 2004 levels.

Supply Growth Estimates

	Domestic Production	Import Volume	Total Supply	Annual Growth %
2002	84,075	24,128	108,203	
2003	88,995	23,269	112,264	3.8%
2004	91,094	26,222	117,316	4.5%
2005	92,296	28,482	120,778	3.0%
2006	93,860	29,396	123,256	2.1%
2007	96,623	29,075	125,698	2.0%
2008	100,528	27,699	128,228	2.0%

Conclusions

Growth, and the industry's ability to supply, will play a critical role in determining consumption levels. Due to promotional efforts, the strengthening economic recovery, and rising cement intensities resulting from the strong competitive position of concrete versus steel, demand for concrete and cement remains strong and is not going to abate anytime soon. PCA projections assume plant operating rates in excess of 98% throughout the forecast horizon. Furthermore, these high utilization rates are supplemented by strong import growth. While shortage conditions may lapse, our forecast suggests that the industry will be running near maximums in terms of domestic production and import capabilities.

In terms of demand, residential construction is expected to perform stronger in 2004 and 2005 than previously anticipated. This reflects PCA's new outlook regarding the path of mortgage rates calling for a delay in rate increases. By 2006, however, single family building starts will record significant and sustained declines lasting through the end of the forecast horizon. Keep in mind, these declines originate from record levels. From a historical perspective, single family starts will remain strong.

While the ongoing economic recovery will be slowed by higher oil prices, nonresidential construction is expected to achieve strong gains in 2005 and beyond. These gains are expected to be supplemented by ongoing increases in cement intensities throughout the nonresidential sector during the 2005-2006 period. Promotional efforts as well as the price of steel relative to concrete account for PCA's optimism with regard to nonresidential intensities. Although nonresidential cement growth rates are slightly slower rates than anticipated in the summer forecast, the 11.9% growth for 2005 and 14.0% growth in 2006 are still quite strong even when considering the depressed base upon which these rates are based.

Cement demand related to public construction activity holds tremendous opportunities in the years ahead. As employment grows, the tax base in each state grows as well. Revenue growth and surpluses have already replaced the dismal atmosphere of sustained deficits in many states. Our estimates indicate that record state surpluses will begin to surface later in the forecast horizon. With the dramatic reversal of fiscal fortunes at the state level, public construction activity will reflect significant and sustained gains. Pent-up construction demand at the state level is expected to be released beginning in late 2005. Furthermore, for fiscal 2006 a new TEA is expected to be in place at a level of $299 billion – compared to the $280 billion level assumed in the summer forecast. Like the nonresidential sector, cement intensities are expected to reinforce the recovery in the public sector in virtually every category except highways. The expected decline in highway intensities is due to changes in the composition of overall highway and street spending. As the public sector recovers, discretionary highway and streets spending is expected to grow disproportionately. Discretionary highway and streets spending tends to have a higher asphalt and lower cement intensity, thereby lowering overall highway and streets intensity. Overall, PCA expects public cement demand will growth by 3.8% in 2005 and 4.1% in 2006.

PCA's relatively optimistic outlook regarding cement demand is tempered by supply conditions. Through the 2005-2008 period, cement producers have announced plans to add roughly 11 million tons of capacity. PCA assumes these plants will come on line with on time planned start dates. Furthermore, PCA assumes operating rates in excess of 98% throughout the forecast period. Even this ambitious expansion activity will need to be supplemented with strong import levels to meet demand. During the past six months cement suppliers have shown the ability to increase imports in the face of difficult import conditions to meet demand. This effort is expected to endure throughout 2005 and beyond. The current forecast reflects a 12.7% increase in 2004 imports, followed by an 8.6% increase in 2005 imports. As new capacity comes on line in 2006, import levels are expected to moderate and then retreat.

Total United States cement supply is expected to grow by 4.5% in 2004 and 3.0% in 2005. Even with the more aggressive estimates with regard to future import levels contained in the current forecast, more moderate growth in supply materializes in 2006 and beyond – limiting United States consumption growth in those years.

CHALLENGES OF CEMENT PRODUCTION

Duncan Herfort
Aalborg Portland, Denmark

1. INTRODUCTION

Cement Production is driven by the same fundamental forces today as when Portland cement first appeared on the scene a century and a half ago, i.e. the need to minimize production costs, to supply cement fulfilling customer requirements in terms of quality and uniformity for its intended use, and to meet both short and long term demand. On top of this, today's cement producers are redoubling their efforts to lower their consumption of natural resources and reduce CO_2 emissions in response to increasing environmental concerns and regulations. To remain competitive long term planning is needed to maintain and upgrade equipment, to ensure a continuous supply of resources (i.e. raw materials, fuel, and qualified manpower), and to continually adapt to an increasingly sophisticated concrete industry. In taking on these challenges, the cement producer should not loose sight of the fact that cement production is a chemical industry, and that optimum production in terms of minimum energy consumption, minimum impact on the environment, and maximum performance of the finished product, can only be achieved by a detailed understanding of the complex reactions in the kiln. The challenges of cement production, and for the cement industry as a whole, outlined in this paper are grouped under two main headings, 1) the need to lower energy consumption and reduce the impact of production on the environment, and 2) the need increase the competitiveness of cement based building materials.

2. LOWER ENERGY CONSUMPTION AND REDUCED IMPACT ON THE ENVIRONMENT

Cement production has always been an energy intensive industry with the cost of fuel for clinker production representing the largest single variable cost of production. The cement industry has therefore invested heavily in new technology over the last 50 years with most producers nowadays operating multi-stage, pre-calciner kilns and high efficiency separator grinding systems. Whilst the technological advances in kiln and grinding technology continue, the return on capitol investment is becoming less and less attractive as a means of reducing energy consumption and increasing output. This is illustrated in figure 1 for the German cement industry where reduction in the fuel consumption/tonne clinker is at a much slower pace now than it was in the 1950's and 60's, with similar trends in other industrialised counties. Emphasis has therefore shifted towards the increased replacement of fossil fuels by alternative CO_2 neutral fuels, and the replacement of clinker by supplementary cementitious materials, SCMs, such as granulated blast furnace slag, fly ash, and natural pozzolans. However, although SCMs are under-utilised in many parts of the world, supplies are limited. In the US, for example, production of blast furnace slag is less than 10% of the annual cement consumption, and of this only 20% is used as a SCM [1]. The supply of limestone on the other hand, which is now also recognized to have cementitious properties, is essentially limitless. Apart from the increased use of alternative fuels, developments in clinker production are likely to focus on the increased use of mineralization of clinker production, reduced C_3S contents, and development of clinker better suited to SCMs. Examples of some of these developments are given in the following.

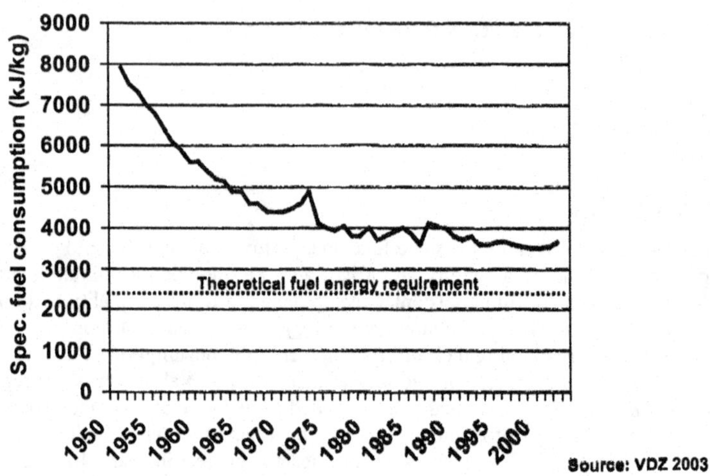

Figure 1. Specific fuel consumption (per kg clinker) in the German Cement Industry since 1950 [2]

2.1 Mineralised clinker

All Portland cement clinkers contain impurities and are therefore mineralised. Impurities enhance the formation of the clinker minerals (mineralization) in two ways, 1) by lowering the temperature of melt formation below the invariant temperature of 1338°C for the C-A-S-F system, and 2) by substitution in the clinker phases which increases the configurational entropy and reduces the free energy at relative low concentrations. All cement producers wittingly or unwittingly mineralise their clinker. Of course, to improve clinker "burnability", generally defined as the ease of alite formation, it is necessary to mineralise the alite phase rather than belite. This is achieved by impurities which are preferentially incorporated in alite. Fluorides and chlorides, for example, are known to enhance mineralization of alite, whereas sulphate is known to stabilise or mineralize belite [3,4,5]. The mechanism of mineralization by reducing Gibbs' Free Energy is illustrated in figure 2. The negative effect of sulfates which stabilise the belite phase is effectively compensated for by adding relative small amounts of fluoride.

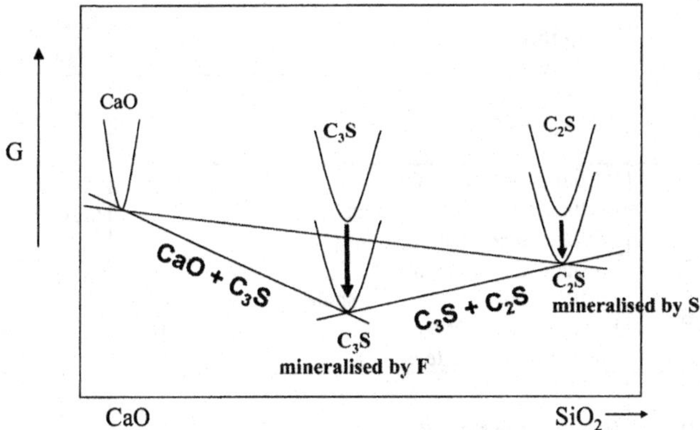

Figure 2. Relative free energies of clinker phases at 1200°C below which C_3S is normally unstable. Mineralization of C_2S on its own destabilises C_3S whilst even small contents of F which is preferentially incorporated in the C_3S phase more than compensate for this by lowering the free energy of C_3S [3].

Sulfate is incorporated in the belite phase by a coupled substitution of one S^{6+} ion and two Al^{3+} ions, for three Si^{4+} ions on the tetrahedral site which significantly increases the hydraulic reactivity of belite. This can be utilized to produce belite rich cements in which 28 day strengths are as high, or higher, than obtained from conventional high alite cements [6].

2.2 Supplementary cementious materials

In Europe "EN 197 CEM II" [7] cements, which contain up to 35% replacement (by GBFS, silica fume, fly ash, natural pozzolans, burnt shale or limestone) account for slightly more than half of the total production of Portland cement. The type III GBFS cements can contain up to 95% slag. With the degree of utilisation of secondary SCMs such as fly ash and GBFS being close to saturation point in Western Europe, limestone is becoming an increasingly important primary source of replacement material, owing to its unlimited availability, and technical benefits in terms of performance, such as reduced water demand and improved concrete consistency. When used with conventional clinker, limestone has a low 28 day activity factor of approximately 0.2, so that limestone filler cements are usually classified as low strength cements (EN197 strength class of 32.5). More recently, however, limestone fillers have been shown to be particularly reactive with mineralised clinker resulting in activity factors close to unity for both early and 28 day strengths as shown in figure 4 [8].

Type		Designation	% clinker replacement
CEM I		Portland cement	0-5
CEM II	A	Portland-slag cement Portland-silica fume cement Portland-pozzolana cement Portland-fly ash cement Portland-burnt shale cement Portland-limestone cement Portland-composite cement	6-20
	B		21-35
CEM III	A	Blastfurnace cement	36-65
	B		66-80
	C		81-95
CEM IV	A	Pozzolanic cement	11-35
	B		36-55
CEM V	A	Composite cement	36-60
	B		62-80

Table 1. Simplified table showing common cement types in EN 197 [7].

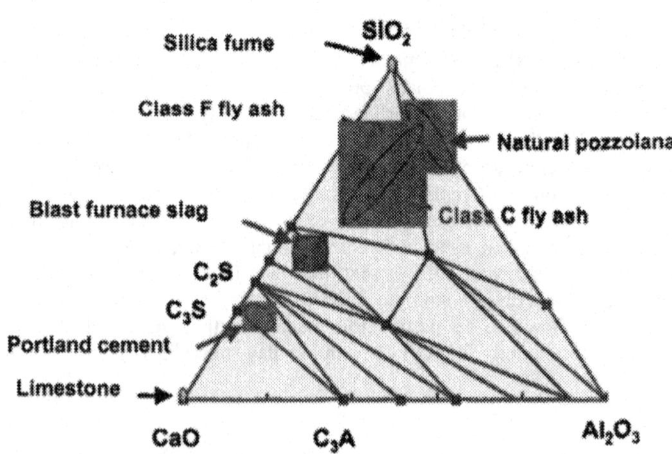

Figure 3. Composition of supplementary cementitious materials. The higher silica contents of the SCMs result in lower Portlandite contents and higher C-S-H contents after hydration. Since C-S-H is a low density phase it fills more space resulting in lower residual porosity, higher strengths and superior durability.

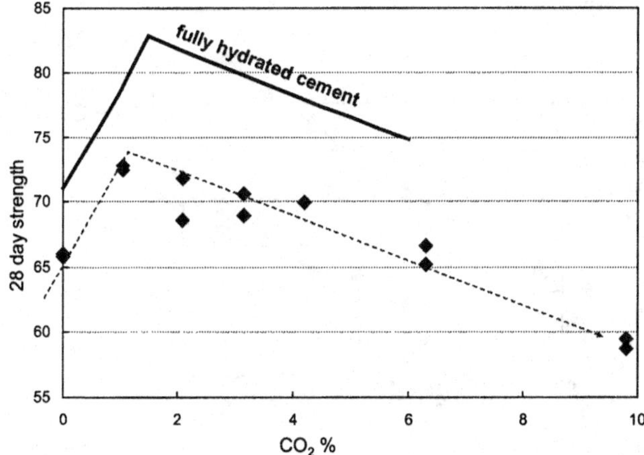

Figure 4. 28 day strengths of cement produced from mineralised clinker at increasing contents of limestone filler, and constant fineness of the clinker fraction. Approximately 15 % limestone addition (just over 6% CO_2) results in comparable 28 day strength compared to the cement without limestone. The strengths are in good agreement with the theoretical strengths of the fully hydrated cements predicted from volumetric calculations [9]. The sharp optimum is not an artefact, but the result of the rapid release of alumina from the ferrite and silicate phases in the mineralised clinker which reacts with the limestone to form the "monocarbonate" AFm phase.

One concern which may limit the use of limestone filler cements in the future is risk of thaumasite formation resulting from external sulphate attack. However, whilst there is little doubt that at sufficiently high sulphate contents limestone will react under cool, water saturated conditions to form thaumasite, the kinetics are poorly understood. Work based on a more systematic, scientific approach, backed up by realistic performance testing, is needed before limestone filler cements can be ruled out in sulphate rich environments.

2.3 Alternative fuels

Alternative fuels are being increasingly used to reduce the cost of production and to lower CO_2 emissions, with Germany leading the way in Europe as shown in figures 5 and 6. Plastics and meat and bone meal make up the bulk of the alternative fuels used today in northern Europe, and because of their high chloride contents by-passes at the back end of the kiln are usually needed to maintain a chloride content of less than 0.1% in the cement (European limit for all "common cements"). The high contents of phosphorous present in meat and bone meal have been reported to stabilise belite [11], but since significant amounts of P_2O_5 are also incorporated in the alite phase this is unlikely to be due to a preferential lowering of the belite phase's free energy as described above for sulphur. Instead, the effect, at least at normal contents of $< 1\%$ P_2O_5 in the clinker, is more likely the result of a coupled substitution of one P^{5+} ion and one Al^{3+} ion for two Si^{4+} ions on the tetrahedral site in both silicate phases, which simply increases the overall content of C_2S at the expense of C_3S, rather than any thermodynamic effect per se. This effect can be compensated for by increasing the LSF of the raw feed in order to maintain constant early strengths by targeting a constant alite content rather than a constant Bogue C_3S content [3].

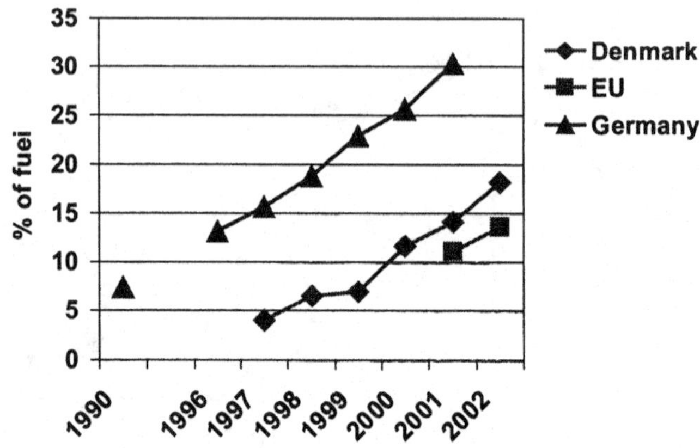

Fig 5. Alternative fuels in clinker production [10].

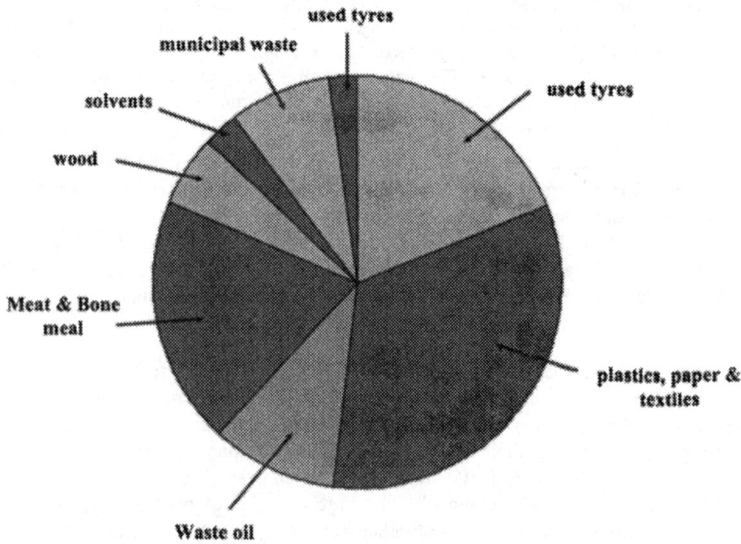

Figure 6. Relative amounts of alternative fuels used for clinker production in Germany [10]

A major obstacle to the replacement of traditional fossil fuels by alternative fuels has been the view among local communities and NGOs that emissions of trace elements and POPs (persistent organic pollutants, mainly dioxins and furins) are increased. Environmental organizations such as Greenpeace, the World Fund for Nature (WFN), the International POPs Elimination Network (IPEN), and the United Nations Industrial Development Organization (UNIDO) have expressed strong views against the use of cement kilns for waste management purposes [12]. This, despite extensive data that fails to show any significant increase in emissions, even of the most volatile elements such as Hg [13,14]. Concerns have also been raised concerning the leaching from concrete of heavy metals originating from alternative fuels or raw materials in which unrealistic test methods for determining the degree of leaching such as the Dutch availability test (NEN 7341) have been cited. However, more appropriate methods such as the Dutch trough method (NEN 7345), which more accurately reproduces realistic conditions, show that for most metals, only 0.001% of the total content present in the concrete is leached after 100 years [15].

2.4 Equipment optimisation

Most advances in kiln technology involve improvements on the existing pre-calciner kiln and grate coolers which remain the back-bone of the industry. A notable example of this is the development of multi-chamber calciners for NOx reduction. This involves a reduction chamber to reduce the formation of NOx gasses from the fuel, and a subsequent oxidation chamber to re-oxidize CO [16].

In grinding technology, more efficient vertical roller mills have largely replaced ball mills for coal and raw meal grinding. Ball mills are still widely used for cement grinding in conjunction with high performance classifiers. Vertical roller mills can result in significant reductions in the power needed for cement grinding but generally lead to more narrow size distributions of the cement resulting in increased water demand and tendency to bleed. This may be resolved in the future by inter-grinding with harder (e.g. GBFS), or softer (e.g. limestone) SCMs, or separate grinding to different surface areas before subsequent blending.

3. INCREASED COMPETITIVENESS OF CEMENT BASED BUILDING MATERIALS

In order to effectively pass on added value to its customers it is important for the cement producer to meet increasingly narrow specifications in an increasingly sophisticated market, and in many cases to supply tailor-made cements for specific applications.

The cement producer can directly add value to the concrete industry by

- reducing the environmental impact of cement (which is factored into the life cycle analysis of the end product), and
- by developing smarter multifunctional cements

and by supplying fit for purpose cements at higher levels of consistency in terms of

- quality
- compatibility with admixtures
- reliability of supply

Increasingly sophisticated concrete production requires very consistent cement quality. As an example, polycarboxylate ethers used in the production of self consolidating concrete can be very sensitive to small variations in PSD, C_3A content, and soluble sulfate content. Increased demand for other specific performance characteristics e.g. high early strength, low heat, chloride resistant, sulfate resistant, low alkali, low water soluble alkali, white and coloured cements, etc., mean that any all purpose cement cannot fully meet all of these requirements and producers in the future will be increasingly forced to supply a range of specialized cements for specific applications, even though this requires considerable investment in additional storage and terminal facilities and places additional strain on the organisational complexity of production and the logistics of supply.

More than ever the cement producer needs to work closely with the concrete industry for cement based building materials to remain competitive against other materials such as plastics, steel and timber. One of the major challenges here is to overcome deep seated views that concrete is environmentally unfriendly and lacks innovative potential. The reality is of course very different with numerous life cycle analyses showing concrete to be more environmentally friendly than most other competing materials, with the embodied energy of cement production accounting for a relatively small portion of the total energy involved in the use of concrete buildings and structures [17]. The cement industry is beginning to take a leading role in challenging these views, as exemplified by "NANOCEM" (nanocem.net) which is a recently formed consortium of Europe's leading universities and producers of cement and other suppliers to the concrete industry. The same popular mind set is also a major obstacle in convincing local authorities that the burning of alternative fuels and raw materials, if managed correctly, does not pose a treat to the local environment, and is in fact the best total solution for disposing of many hazardous materials.

4. CONCLUSION

For cement based materials to successfully compete with other building materials in the future, the cement industry needs to continue to reduce its dependency on non-renewable fuels and to maximise the use of supplementary cementitious materials. Equally important is the need to convince policy makers and the public that concrete (and other cement based building materials) is the best total solution for most applications in terms of innovation potential, functionality, and environmental compatibility. Innovation at the individual cement plant is central to these challenges where efforts to minimize energy consumption and CO_2 emissions must go hand in hand with a detailed understanding of how the complex chemical reactions in the kiln ultimately affect cement performance.

REFERENCES

1. U.S. Geological Survey, Mineral Commodity Summaries, January, 2005.
2. VDZ "Activity Report 2001-2003" Verein Deutscher Zementwerke e.V. Verlag Bau+Technik GmbH, Düsseldorf. 172 p.
3. Kerton, P., "Sulphate and Fluoride" International Cement Review, Sept, 2002, pp. 73-78.
4. US patent no. 4042408.
5. Dagen, S, Welqing, C., Qixiu, H.,Huo,M., & Lipiang, Z., "Effect of SO3 on Mineral Formation and Properties of Clinker", 9th Int. Congress on the Chemistry of Cement, New Delhi, 1992, vol. II, pp 322-328.
6. Herfort, D., Holmboe, A.S., Costa, U., Gotti. E. & Grundvig, S., "High Strength, Low Energy Cement Based on Belite Rich Clinker", Proceedings of the 8th Euroseminar on Microscopy Applied to Building Materials, 2001, pp. 139 – 146.
7. European Standard EN 197-1:2000, Composition, specifications and conformity criteria for common cements.

8. Borgholm, H.E., Herfort, D. & Rasmussen, S., "A New Cement Based on Mineralised Clinker", World Cement, August 1995, pp. 27- 33.

9. Nielsen, E.P., Herfort, D. & Geiker., "Phase Equilibria of Hydrated Portland Cement", Cement & Concrete Research, 2005, No.1.

10. VDZ "Activity Report 2001-2003" Verein Deutscher Zementwerke e.V. Verlag Bau+Technik GmbH, Düsseldorf. 172 p. + Aalborg Portland Environmental Reports 2000, 2001, 2002

11. Halicz, L. & Nathan, Y., "The Influence of P_2O_5 on Clinker Reactions", 1983, Cement & Concrete Research, 1983, No.14, pp. 11-18.

12. "Study on POPs Emissions in the Cement Industy", CEMBUREAU Eurobrief No 123, April 2004.

13. Oerter,M., "Effects of the use of secondary fuel on Emissions in clinker production", Proceeedings of the VDZ Congress, 2002, pp. 368-371.

14. "Formation and release of POPs in the cement industry", SINTEF Report for the World Business Council for Sustainable Development, March 31, 2004.

15. "Environmental Compatability of Cement and Concrete", in VDZ "Activity Report 2001-2003" Verein Deutscher Zementwerke e.V. Verlag Bau+Technik GmbH, Düsseldorf, pp. 124-134.

16. Kasch,K.H., Proceeedings of the VDZ Congress, 2002, pp. 212-215.

17. Kuhlmann, K. and Paschmann, H., (1997), "The ecological position of Cement and Concrete", ZKG International, 50(1), pp.1-8.

CEMENTS FOR THE FUTURE

F. P. Glasser
University of Aberdeen, Old Aberdeen, AB24 3UE, Scotland, United Kingdom

ABSTRACT

The driving forces for change are discussed. A range of responses is envisaged ranked according to increasing extents of change. New pozzolanic materials can be developed to supplement Portland cement. Unconventional technologies can be used, for example, to co-produce cement and electricity. The performance lifetime of concrete could be extended by enhancing the durability of concrete construction Organic admixtures can be better exploited. The potential for novel cement types is discussed: sulfoaluminate cements and reactive belite cements are considered to offer the best scope for technical advances and commercialisation.

1. INTRODUCTION

When I was offered this title, I appeared to have a relatively clear field, tempered only by the knowledge that most predictions of the future not only turn out to be wrong but also to look foolish. However, the appearance of a thoughtful paper on the future of industrial cement production by Gartner [1] galvanised me into action, especially as some of my thoughts were pre-empted by its contents! On further reflection, there are still useful things to say. I concentrate first on the drivers of change and then develop possible responses.

2. DRIVERS OF CHANGE

Change rarely occurs spontaneously: it occurs in response to perceived needs opportunities and challenges. What are the drivers of change in the cement industry? Here, I use the term "cement industry" loosely to include not only cement production but also to include the extraction of raw materials for cement and concrete making, as well as those individuals or groups who specify, use, or otherwise have a stake in concrete. This latter group, of stakeholders might for example, include the owner of a dwelling house built on a concrete slab foundation, or whose water supply is provided by a network of concrete dams and pipes.

Most cement today is produced in the developing world. The "traditional" markets, of Europe, North America, Japan etc., have significant but relatively stable production. Thus the bulk of cement production goes into creating an infrastructure. As world standards of prosperity rise, demands multiply for shelter, for potable water for rural and urban use, for construction of schools and hospitals and numerous other infrastructure improvements. This accounts for the impressive annual increases of cement production in the developing world. But developed countries also need to reconstruct their infrastructures in response to population changes and socio-economic pressures. The cement industry, it could be argued, has responded well to these challenges. The product it makes is relatively cheap, reliable and competes well with rival materials. So cement production will continue to increase - not indefinitely, but for the foreseeable future.

But the very success of the industry compels change. Cement production already generates a significant fraction, perhaps 10%, of world carbon dioxide emission. The industry is doubly handicapped by the fact that its principle resource, limestone, has to be decarbonated thus giving rise to carbon dioxide from raw materials as well as from combustion of fuel. Hence the large carbon dioxide outputs. We may argue about the role of carbon dioxide in climate change, but the balance of probability is that a real link exists; regulation and taxation certainly presume that the link exists.

We live in an era where sin- in this case, emission of carbon dioxide- is likely to be punished by taxation. Since cement production occurs mainly at relatively few large installations, production is easily monitored and hence readily taxed. Taxation, or threat of taxation thus becomes an important driver of change. But even without taxation, the rising cost of energy has already prompted considerable energy saving in the cement industry: Gartner estimates as much as 40% saving per ton having been achieved in North America between 1974 and 1994. This gain has resulted largely from replacing old and inefficient equipment with modern plant. Unfortunately the large gain in thermal efficiency does not directly translate into reduction of carbon dioxide emissions because the need to decarbonate the raw materials remains unchanged. Moreover, and looking to the future, scope for further gains in efficiency increasingly encounter thermodynamic constraints. We have, very nearly, realised the limits of energy efficiency of conventional plant. Thus more radical measures are needed to ensure future gains.

3. SUPPLEMENTARY CEMENT MATERIALS

The use of supplementary materials is well established. Slag and coal combustion fly ash are arguably the best known and have been described in countless publications, reports and standards. In favourable cases, these so- called blended cements may have superior properties to plain cement concretes. However utilisation of these supplementary materials still lags: it approaches 100% only in a few markets. Globally, and even allowing for the poor distribution relative to demand of the relevant supplementary materials, more could be done to replace cement. However much of the enabling research has been done and better utilisation will follow: for the future, research needs to focus on hitherto untapped resources.

4. NOVEL POZZOLANIC RESOURCES AS CEMENT SUPPLEMENTS

Supplies of traditional pozzolanic materials are, as noted, ill distributed and, moreover, pressures on industries producing fly ash and slag will tend to reduce somewhat supplies of these materials. For example, better pre-processing of iron ore and pre-reduction, perhaps occurring close to source will lessen the amount of slag needed to cleanse the metal in an iron blast furnace.

Where should the search for novel pozzolanic materials begin? I suggest two such areas. One is utilise a greater range of clays than has hitherto been used and the second lies in natural resources of zeolitic and glassy rocks.

Clay, in the form of kaolin, is known to form the basis of an excellent pozzolan. Careful dehydroxylation of the precursor, kaolin, leads to loss of water and formation of a poorly crystallised product that preserves some structural elements of the precursor kaolin; on that account it is termed "metakaolin". Several years ago, Dow [3] examined metakaolin and found that one particular metakaolin performed extremely well in blends made with cement, portlandite and other pozzolanic materials. We were initially at a loss to explain the beneficial action of this particular metakaolin. But research showed that this "metakaolin" also contained a significant amount of illite. Illite in this context was a degraded potassium mica (muscovite). It gave an extremely weak and diffuse X- ray diffraction pattern and was not at first noticed, but in fact, comprised about 10% of the bulk sample. In the course of converting kaolin to metakaolin, its illite content had also been activated so, upon mixing with a source of calcium hydroxide, the degraded illite underwent ion exchange and reaction with liberation of potassium, effectively as the hydroxide. The KOH thus released activated other substances, eg, slag or fly ash.

An extensive literature exists on alkali activated cementitious formulations and KOH is often found to perform better in this application than NaOH. Yet its expense has made the economics of KOH activation impractical. But thermal activation of illite provides a convenient slow release form of KOH at low cost. Because potassium is released slowly, the wet pastes are not excessively caustic, as occurs when KOH has to be added to the mix water. Moreover, as reaction progresses, the aluminosilicate reaction products bind K so that it may not be necessary to use an alkali-resistant aggregate when formulating concrete products.

This example shows that there are many combinations of potential pozzolanic resources remaining to be exploited. Since many of these resources are not well-distributed, it may be necessary to develop a range of products, each suited to locally available raw materials. Experience of coal combustion fly ash, itself a highly variable product with respect to composition and mineralogy, shows that the industry is sufficiently robust to cope. The fly ash experience is thus relevant to exploiting highly variable natural resources, e.g. zeolite and glassy deposits as pozzolans. Their intrinsic variability is best overcome by selective extraction, beneficiation and blending, to assure a constant reactivity of the bulk product.

5. EXPLOITING UNCONVENTIONAL TECHNOLOGIES

Unconventional technologies appear in a variety of guises. We could initiate change in a variety of ways: for example by seeking to change the way in which we make cement or to modify the properties of existing production or we could change the nature of the product. The more radical the change, the slower it will be to achieve: "new" cements are regarded with general suspicion and it must be admitted that problems have arisen in the past with so-called "wonder" materials. Nevertheless cements and concretes are used in a range of applications, structural as well as non-structural and new cements may first be introduced in a range on non-structural or at least not very demanding structural applications. Certainly all options for change should be considered at least in the first instance. I offer selected examples considering in the first instance improvements to conventional materials but subsequently consider more radical changes.

6. ENHANCED DURABILITY

The service lifetime of a concrete structure is highly dependent on exposure conditions as well as design, formulation and execution. Benign conditions of exposure rarely present problems. But severe environments present challenges. As the writer and others have stressed, "durability" cannot be defined: it is an abstract concept [3]. However abstract words have *attributes* that can be defined and it is in this area, defining and quantifying and implementing the relevant set of attributes in relation to service conditions, where we must progress. Increasing the service lifetime of concrete in aggressive environments is one of the biggest improvements we could achieve to better the energy efficiency of construction. Additional socio-economic benefits accruing from improved durability have to be quantified on a case-by-case basis; for example, disruption and economic loss arising from replacement or repair of major components of a transport network attach a high degree of benefit to long service life. For this reason, many new key infrastructure investments have specified minimum guaranteed performance lifetimes. Whether these targets can be achieved with existing technology not entailing excessive cost remains an important point.

An example suffices. This series of meetings has in the past discussed the sulfate resistance of concrete and a number of summary publications of the literature have also been made subsequently [4-6]. I will not attempt to recapitulate the evidence and arguments but to an outside observer it might seem striking that, although the existing standards for sulfate resistance were generally agreed to be inadequate and insufficient, and much new research on the causes of sulfate attack has been reported, no attempts were evident to utilise these (and other) research findings to revise the standards. Several years later the writer does not see significant progress having been made. Thus we are condemned for the foreseeable future to use empirical tests and test data, often much out of date and lacking proper experimental design, with which to specify sulfate-resistant concretes.

The writer has had the opportunity to participate in plans for nuclear waste disposal in the course of which the durability of several classes of material were *calculated*, e.g., steel, glass and ceramics. Comparing the sophisticated approaches to predicting the durability of these materials, the selection of concretes and cements for durability, as well as verification of performance, seems to lag. We could do much better by making a sustained effort with the application of the methods of modern materials science.

So a recommendation is that a totally new approach to durability be made, especially in respect of chemically-induced deterioration. Many papers highlight the way forward. Equilibrium, kinetics and mechanics have to be integrated into a single model. I do not suggest that such a model will emerge full-blown: progress may well be uneven and an overarching model, embracing all aspects of durability, may well not be attained for decades. But a start needs to be made if we are to quantify durability in terms of accepted physico-chemical parameters.

7. UNCONVENTIONAL MANUFACTURE OF CEMENT

Although the rotary kiln is a satisfactory way of making cement, it produces much low-grade thermal energy, which is difficult to recover. Alternative methods of cement making should therefore be investigated. One such method involves co-production of cement and electricity. In this process, shown schematically in Fig.1, the ash content of coal is utilised to supply some of the cement components but is supplemented by mixing with other components, e.g., limestone, to achieve a proper mix design for a cement clinker. The mixture is burnt, perhaps in a fluidised bed to permit longer residence time than is achieved in conventional combustion. The ash is collected and ground to furnish cement. If the offgas is clean, it can be used to drive a gas turbine in tandem with a second stage steam cycle; such arrangements give very efficient electrical conversion. If, however, the hot gas stream contains much particulate matter, it may be necessary to rely on a steam cycle as is done in conventional coal fired plant. Some of the electricity thus produced is used internally, for example to grind the cement. Co-production with an electrical output avoids some of the difficulties associated with previous two-product processes. Thus Gartner [1] refers to a process, not now in commercial use, whereby gypsum was used to make cement: the second product, sulphuric acid, was commercial but in practise it was difficult to match supply and demand for the two quite different chemical products, quantities of which were fixed by stoichiometry.

Unconventional technologies may well be amenable to making non-Portland cement types. For example, powdered gypsum or anhydrite is sometimes injected into coal combustion systems to reduce NO_x emissions. This suggests that the co-production process might be amenable to production of other cement types: for example of calcium sulfoaluminate-based cements.

8. MODIFICATION OF CEMENT PROPERTIES

The development of plasticisers has led to a range of low water content concretes having enhanced durability characteristics. The enhancement is associated with low permeability, but without the penalty normally associated with low w/s ratios, of poor flow and compaction of fresh concrete. First generation plasticisers were mainly industrial by-products, unwanted and hence cheap. Now, with the general acceptance that organic admixes can be beneficial to concrete properties, the industry has for the first time achieved the confidence to consider tailoring organic molecules for specific purposes. The tailoring concept is implemented by molecular design. Molecular design in this application is still in its infancy but the area of molecular design and of molecular interactions with cement substances is one of immense promise. Empirical work enabled Birchall and colleagues to obtain very high compressive and flexural strengths [7]. While these early formulations were not resistant to hydrolysis in wet service, resulting in loss of strength, the research did usefully highlight how cement might be given a range of novel and hitherto unexpected properties. I am not necessarily advocating a return to the formulations described in [7], but a valuable conclusion from these studies was to demonstrate that at high pH, bonding developed between water-dispersible PVA and Al from the hydrated paste components: thus organomineral bonds can greatly modify and in some cases improve the efficiency of cement materials in structural applications. These and other interactions, perhaps assisted by intercalation and dispersion, could lead to development of new classes of materials, based on organo-mineral composites.

Other and unanticipated applications await development and I turn to processing for my next example. The performance of concrete in many products would be much improved if we could develop a cheap, reliable and controlled method of forming functionally graded concretes. Many application areas for such materials could be cited: for example, concrete slabs or panels having lightweight, porous cores, for thermal and acoustic insulation, but with dense and impermeable surfaces for better load bearing and weathering properties. We are probably all aware that the surface of concrete, the so-called "covercrete" differs in properties from bulk concrete, but have yet to control and manipulate these differences constructively, to achieve improved composite properties.

As a further example of how concrete properties can be manipulated, the writer and colleagues have been working to develop "breathable concretes". The concept is straightforward: a sealed construction is kept under slight negative pressure, *ca* 10 Pa. Incoming air flows through an air-permeable concrete. Assuming a cold climate, incoming air is warmed, thus improving the apparent insulation value of the concrete. If required, humidity can also be controlled. Used as panels or wall sections, construction is greatly simplified; separate insulating layers are not required. Innovating concretes to meet the specifications for ventilation, while at the same time retaining adequate structural strength, has been a challenge!

10. CHANGES AFFECTING THE NATURE OF PORTLAND CEMENT

Until now, the range of formulations discussed involves Portland cement as presently made: its properties are modified after it is made. However a range of modified cements are attractive. Gartner [1] reviews perspectives and progress in belite cements. Briefly, their advantage lies in lowering the chemical CaO content of clinker, resulting in less carbon dioxide emission per unit of output. The high alite content of modern cement would be reduced while that of belite, essentially dicalcium silicate, would be enhanced. The disadvantage is that belite-rich clinkers are more difficult to grind and, at comparable surface area, their strength gain is less rapid than for conventional alite-rich clinkers.

The reactivity of belite in commercial Portland cement varies widely from one clinker to another. This arises, I conjecture, partly as the result of the complex thermal history of belite. At the high temperatures obtaining in the kiln, belite is present as the alpha phase. The structure of alpha belite is salt- like, in this case isostructural with that of high potassium sulfate. Its comparatively open structure enables it to form solid solutions with many of the abundant oxides in the clinker batch. As the clinker is cooled, the belite phase undergoes a complex series of phase transformations. These transformations lead to phases having progressively lower symmetry than the precursor. In the course of transformation, two types of event occur: twinning, initially with coherent twin boundaries, and exsolution. As Fukuda and colleagues have shown [8-9], exsolution can in some cases result in partial remelting within the cooling crystal: with continuous withdrawal of heat, a single homogeneous solid spontaneously transforms to another solid and melt. The exsolved products are morphologically aligned by crystallographic controls operating during cooling. Although it was not possible to give a full explanation of the origin of the complex internal structure of clinker albeit at the time. Insley [10] recognised from microscopic studies of cooled clinkers that different varieties of belite existed, characterised by colour, type and spacing of twin lamellae.

At high temperature, or in the temperature range at which these processes initiate, the lattices of host and exsolved phases (except for a melt phase) are coherent. However, with further cooling, and owing to different dilational coefficients of the phases involved (or formed, as the exsolved melt eventually freezes with crystallisation), strain energy accumulates at phase boundaries. This, in turn, affects the reactivity of the belite. The effect can significantly alter the bulk thermodynamic properties and, in turn, affect the kinetics of its subsequent hydration reactions. Because of the frequency with which internal lattice misfits occur in many belites, the reticular spacing between exsolution lamellae is often on the order of a wavelength of visible light or less. Thus misfits occur in sufficient population to alter bulk properties: in this instance, rate of hydration. It is therefore concluded that the search for "reactive belite" should not rely entirely on empirical formulations and heat treatments but should systematically integrate characteristics of the fine structure of belite coupled with its thermal history and with appropriate measures of reactivity: but at this point in time it is not certain what measurements of fine structure need to be made, how best to make the measurements and how to select appropriate measures of reactivity: this is the goal of research.

11. SULFOALUMINATE CEMENTS

Since I was first introduced to the Chinese product, I have been fascinated by the properties and potential of these cements and could not be deflected from mention of sulfoaluminate cements in of my presentation! But, to start at the beginning, these cements are made using limestone, bauxite and gypsum. A normal rotary kiln can be used for their production although the flame and burning zone temperatures befit from adjustment because peak burning temperatures are only ~1325°C [11] i.e., rather less than for Portland cement. However the main savings in energy and carbon dioxide reduction arise because of the lower calcium content of sulfoaluminate cement: about half that of Portland cement. As might be expected, the clinker mineralogy differs significantly from both those of Portland or high alumina cement clinkers. Empirical and practical development shows that the optimum target mineral assemblage comprises: belite, calcium sulfoaluminate, ferrite and anhydrite. Calcium sulfoaluminate, Kline`s compound, is $4CaO. 3Al_2O_3.SO_3$ with perhaps minor iron substitution for Al. Free lime is essentially absent in well-made clinkers. Compared with Portland cement clinker, sulfoaluminate clinkers do not contain alite and, upon hydration, do not develop free portlandite. Since sulfoaluminate clinkers are usually interground with 10-20% gypsum pastes develop mainly ettringite in the course of hydration.

A further saving of energy results at the formulation stage: as a consequence of adding more gypsum, the energy rich clinker is diluted. Grinding is done to approximately the same surface areas used for Portland cement but the softer sulfoaluminate clinkers are more readily ground than Portland clinkers. The low sulfate content of the clinker, 8-10%, is thus supplemented by gypsum.

Sulfoaluminate cements are used like Portland cement: sand and coarse aggregates are added to formulate, respectively, mortars and concrete. The water demand is similar to Portland cement but, on account of rather short times to initial and final set, it is normal practice to use retarding plasticisers to achieve low water, high strength concretes. Once set occurs, strength gain is rapid. The writer has visited a Chinese pre-cast factory and seen beams intended for highway structures, cured overnight at slightly elevated temperatures and moved to site for placement within 24-48 hours, by which time they have achieved >40MPa compressive strength. The concrete has good dimensional stability and, by adjusting upwards the gypsum addition at the mill, it could, if desired, be made slightly expansive. However, normal production is intended to be dimensionally neutral.

The matrix of the sulfoaluminate paste consists of ettringite and C-S-H, the latter derived mainly from belite: C-S-H thus formed has a Ca/Si ratio close to 1.5. Any extra "lime" is consumed in the formation of other hydrates, e.g. ettringite, so free portlandite is absent. Concerns have been expressed about the future performance of these materials owing to their low internal pH and potential susceptibility to carbonation. The following partly address these concerns.

We recently examined sulfoaluminate concretes cored from various structures and report the results [12]. The concretes were made years ago without the intention that they would subsequently be sampled: exposures ranged in duration between about 4 and 20 years. The exposure conditions were uncontrolled and Portland cement concretes were not available as controls. The original concrete specification was available but no corresponding quality assurance records were kept. Nevertheless all the concretes were still in service and visual examination of the relevant structures did not reveal any deterioration apart from occasional defects in workmanship. When depths of carbonation were determined for concrete exposed to rain and normal atmospheres, carbonation appeared to have proceeded at a rate similar to Portland cement concrete of comparable quality.

What was of particular interest was a concrete pipe exposed to seawater in an intertidal zone for several years. We report on the condition of the pipe in [12]. The pipe was steel reinforced with only a few mm of cover in places. The pipe had been formed by centrifugal casting; the effective water: cement ratio was affected by centrugation and in all probability was reduced to about 0.25 in the course of centrifugation. Despite the thin cover, the steel wire reinforcement remained bright and uncorroded.

An important characteristic of sulfoaluminate cement clinkers containing much Kline`s compound and gypsum is the high chemical water demand required for their complete hydration [13, 14]. The cement reported in this study probably had a chemical water demand, expressed as a w/c ratio, in the range 0.50 to 0.55. Not surprisingly, given the lowering of the w/c ratio by centrifugation, clinker was abundant. What was arguably more surprising was the persistence of unhydrated clinker to within 0.5 to 0.8 mm of the surface of the concrete; despite tidal immersion with suction, insufficient water penetration occurred to initiate and continue hydration. The matrix pH, about 11.5, was adequate to protect steel despite deep penetration of chloride. Because the paste is essentially sulfate saturated, it does not react significantly with sea water sulfate in service conditions.

While additional research and verification is required, with comparative trials of Portland and sulfoaluminate concretes, experience obtained thus far points to the excellent resistance of sulfoaluminate cement to sulfate attack in general as well as to combined magnesium, sulfate and chloride attack, as occurs in sea water. In this respect, sulfoaluminate cements resemble "supersulfated" (Portland) cements, a product only rarely encountered today but one that was widely used in the past because it was known to resist seawater attack. Supersulfated cements were formulated from iron blast furnace slag, activated by gypsum and Portland cement. Despite their durability, they had the reputation of being difficult to handle: they were prone to false set and even if false set could be avoided, gained strength only slowly. Sulfoaluminate cements would seem to offer a considerable improvement in terms of having well controlled and consistent set. Moreover, empirical evidence is that the cover concrete protects embedded steel against chloride attack.

Thus many aspects of sulfoaluminate cements deserve exploration. Resources for their production are not present everywhere and it is likely that they are most likely to attract favourable interest in non-glaciated regions, such that deep weathering has preserved aluminous deposits. It is noteworthy that high-grade bauxite is not required for their production: the need also to form clinker belite enables siliceous resources to be used that would be rejected for Bayer process alumina.

SUMMARY

The "road map" to the future is complex: no single route exists, but the destination is clear; to develop cementitious products having improved functionality and performance but to lower emissions in the course of production. The potential of Portland cement, as we know it today, can be enhanced by improved durability and as the matrix forming component of organomineral composites. Its composition can be further modified by developed by novel pozzolanic admixtures and it can be produced by unconventional processes. However, more drastic changes are also considered, for example the development of reactive belites and in the recent development of calcium sulfoaluminate cements. To sustain these developments, a materials science orientated approach is needed; present day empiricism has, served us well in the past but has, very nearly, reached its sustainable limits.

REFERENCES

1. E. Gartner "Industrially interesting approaches to "low-CO_2" cements". Cement and Concrete Research, 34, 1489 – 1498 (2004).
2. C. Dow and F.P. Glasser "Alkali releases from crushed minerals and thermally-activated constitutents of metakaolin". Advances in Cement Research, 15, 137-144 (2003).
3. F.P.Glasser "The very long terms performance of cement and concrete", pp 159-166 in Concrete Technology for a Sustainable Development in the 21st Century. O.E. Gjorv and K. Sakai (Eds). E&F Spon (London & New York) (2000). ISBN 0-419-25060-3.
4. J.Skalny, J.Marchand and I.Odler (Eds). "Sulfate attack on concrete". Modern Concrete Technology, 10, Spon. Press (London & New York) (2002). ISBN 0-419-24550-2.
5. K.Scrivener and J.Skalny "Internal sulfate attack and delay ettringite formation". Proceedings of the Rilen TC-186-ISA Workshop, Rilem, (Paris) (2002). ISBN 2-912143-44-6.
6. P.Damidot, S.J. Barnett, F.P.Glasser and D.E.Macphee "Investigation of the CaO-Al_2O_3-$CaSO_4$-$CaCO_3$ system at 25°C by thermodynamic calculation". Advances in Cement Research, 16, 69-79 (2004).
7. J.D Birchall, A.J. Howard and K.Kendall. Nature (London), 289, 388-389 (1981).
8. F.P.Glasser "Advances in cement clinkering", Chapter 3.4, pp 331-368 in "Innovations in Portland Cement Manufacturing". J.I Bhatty, F. Miller and S.H. Kosmatka (Eds). Portland Cement Association, Skoke, IL. (2004). ISBN 0-89312-234-3.
9. K.Fukuda "Recent progress in crystal chemistry of belite: intracrystalline microtextures induced by phase transformations and applications of remelting reactions to improvement of hydration reactivity". Journal of the Ceramic Society of Japan, 109, 543-548 (2001).
10. H.Insley "Structural characteristics of some constituents of Portland cement clinker". Journal of Research, National Bureau of Standards (U.S), 17, Research paper 917, 353-361 (1936).
11. O.Andac and F.P.Glasser "Polymorphism of calcium sulfoaluminate, $Ca_4Al_6O_{13}.SO_3$ and its solid solutions". Advances in Cement Research, 6, 57-60 (1994).
12. L.Zhang and F.P.Glasser "Investigation of the microstructure and carbonation of CSA-based concretes removed from service". Advances in Cement Research (in press 2004).
13. L.Zhang and F.P.Glasser "High performance cement matrices based on calcium-sulfoaluminate-belite compostions". Cement and Concrete Research, 31, 1881-1886 (2001).
14. F.P.Glasser and L.Zhang "Calculation of the chemical water demand for hydration of calcium sulfoaluminate cement". Proceedings of the 4th BISCC, Beijing, PRC, Volume 3, 38-44 (1998).

NANOCEM
An Industrial - Academic Partnership for Fundamental Research on Cementitious Materials

Karen Scrivener
Ecole Polytechnique Federale de Lausanne, Switzerland

This article describes the creation of NANOCEM – a unique Industrial / Academic partnership for fundamental research on cementitious materials with the vision to form an integrated research and education environment to generate and disseminate basic knowledge on the nano- and micro-scale phenomena which govern the macroscopic performance of cementitious materials.

Despite being the most used material in the world the fundamental mechanisms underlying the behaviour of cement remain relatively poorly understood, due to its chemical and physical complexity. Processes occurring during the first seconds and minutes of reaction with water affect the microstructure and consequently the long-term performance of a structure, whose lifetime may extend for a hundred or more years.

These processes – cementitious bonding, material transport, etc. – which control the performance of cementitious materials occur at the nanoscale, but research has largely been limited to the macroscopic level. The consequent lack of detailed understanding results in slow and incremental development. The need to test properties iteratively on the macro level is a major barrier to innovation and to the full exploitation of the potential of cementitious materials, which are widely available; low energy (Figure 1)[1] and non-toxic.

COMPARATIVE ENERGY COSTS OF MATERIALS
(MJ per ton)

CONCRETE	**600 – 800**
WOOD	
Cut wood	**~500**
Multilayer board	**~4000**
GLASS	**15700**
STEEL	**21000**
(from scrap)	**11000**
ALUMINIUM	**164000**
(recycled)	**18000**
PLASTICS	
HDPE	**81000**

[1] Comparative energy costs from: P.Eyerer and H-W Reinhardt ""Ökologische Bilanzierung von Baustoffen und Gebäuden", Basel : Birkhöuser-verlag, 2000) ["GaBi-Projekt"- final report]; "Ökobilanzen Holz" der Dt. Ges. f. Holzforschung, München, April 1997); H.-W. Reinhardt et al., "Sachstandbericht Nachhaltig Bauen mit Beton" .- Berlin: Beuth, 2001. - (DAfStb: Schriftenreihe 521) ; H. Glässer, H.K. Gründler, H. J.-F. Heitz, "Ökologische Betrachtung und Bilanzierung von Wärmedämmglas", Ergbnisbetrachtung (interne Studie der VEGLA GmbH), VRGLA GmBh 1996, Aachen.

At present, any structure involving innovative high performance cementitious materials necessitates protracted testing, which still leaves uncertainties regarding the extrapolation of results from short-term laboratory tests to the performance of materials designed to last for many decades in highly variable natural environments. For example, the Öresund link between Denmark and Sweden required a service life of 100 years from the concrete without any major repair works. The testing programmes started 3 years before the answers were required and are still continuing. In spite of this multi-year test programme, the knowledge to predict long-term performance is lacking. This situation can only be improved through the generation of fundamental knowledge, which can lead to high performance materials being created and selected by design, rather than by trial and error.

Advances in other structural materials, e.g., metals, ceramics, have been achieved by extending the scale of investigation from *macroscopic* to *microscopic* (μm) level and, ultimately, to the *nanoscale* (nm) level. For example, no modern account of the theory and practice of metallurgy would be complete without an explanation of how nanoscale phenomena such as defects affect material properties. Of course cement materials differ fundamentally from metals: direct technology transfer from metals is not possible. Nevertheless *microstructural* studies of cement conducted over the past few decades have greatly influenced the way cements are used. The development of new techniques to characterise materials at the nanoscale has provided a unique opportunity to go further in our understanding of cementitious materials.

The NANOCEM consortium has arisen from the growing realisation of the industrial partners that present developments in the marketplace are incremental and will become asymptotic without improvement in understanding the fundamental micro and nano scale mechanisms governing macroscopic behaviour. The research needed to provide this cannot be financed by any single company. Up to the early 1980s basic research on cement and concrete was carried out at large government or industry funded labs – e.g. Building Research Establishment (BRE), Cement and Concrete Association (C&CA) in the UK; National Bureau of Standards (NBS) and Portland Cement Association (PCA) in USA. Despite the lack of sophisticated modern characterisation techniques, much important research was done in these laboratories, which is still referred to today (e.g. Powers and Brownyard, Lerch, Jones, Lea). The economic climate of the 1980s led to the drastic reduction in size and scope of these laboratories.

In the last quarter century there has been a revolution in experimental techniques for the investigation of materials – for example Nuclear magnetic resonance spectroscopy (NMR), Atomic force microscopy (AFM) and progress in electron microscopy, but on the academic side there is an appreciation that the different characterisation methods need to be integrated into a common approach and that no one research group can put together the critical mass of equipment and expertise required.

Within this context, a new European Research Network has been established, comprising 21 academic and 11 industrial partners with an interest in cementitious materials, from 11 European countries. This includes four of the five largest cement producers representing some 20-30% of world cement production. The members of Nanocem collectively have access to a large range of state of the art equipment for the study of cementitious materials, including electron microscopy, magnetic resonance spectroscopy, atomic force microscopy, focussed ion beam techniques, synchrotron tomography etc. The figures illustrate some results possible with these techniques: Figure 2 shows the three dimensional pore structure of a 3 day old cement paste obtained by synchrotron X-ray tomography by the research group at EPFL, Switzerland.

Figure 2

3 dimensional pore structure in 3 day old cement paste imaged by synchrotron X-ray tomography, Emmanuel Gallucci, EPFL, Switzerland

Figure 3 shows the atomic structure of a C-S-H, imaged by Atomic Force Microscopy by the group at the University of Bourgogne, France

Figure 3

Atomic resolution of C-S-H structure (main binding phase of cementitious materials) from Lesnieska University of Bourgogne, France

The industrial partners jointly provide new finance to fund joint research by the academic partners and help to integrate their independent research work through a better understanding

of the key technical questions, answers to which can lead to breakthroughs in the use of cementitious materials. The academic partners each contribute at least one of their independently financed research project with the network as a whole and seek to develop their research portfolios in a synergistic and complimentary way. The consortium seeks to establish a constructive dialogue between the industrial and academic sectors, through focussed workshops to establish the important gaps existing in our knowledge and joint research projects to fill these gaps. The research is at a pre-competitive level allowing collaboration across the industrial sector. The Network co-ordinator is Karen Scrivener, Professor of Construction Materials at EPFL (Ecole Polytechnique Federale de Lausanne), Switzerland. The steering committee consists of five representatives of the industrial partners and five of the academic partners. Full meetings of the consortium take place twice a year

At present the research programme of the network consists of three core projects and 21 partner projects. The subjects chosen for these first core projects illustrate the need for basic research, which can have a wide range of applications:

Hydrate Assemblages Containing C-S-H
At present it is not possible to determine quantitatively the assemblage of hydrate phases which will result from the hydration of a cementitious system, especially when supplementary cementing materials (SCMs) and fillers such as slags, fly ashes, natural pozzolans, limestones and dolomites are included. This project aims to define the compositions and stabilities of the hydrated phase assemblages that are likely to occur over the temperature range 0-50°C. This will allow the densities or porosities of complex hydrated cements to be calculated as a function of degree of reaction, and thus facilitate the design of blended cements by allowing modelling of their mechanical properties and chemical stability. This PhD project will be conducted jointly by the University of Aberdeen, UK (F. Glasser), EMPA in Switzerland (B. Lothenbach), and ESPCI in Paris (H. Zanni).

Pore Structure by NMR
The objective of this project is to refine a comprehensive set of state-of-the-art non-destructive, non-invasive instrumental techniques capable of fully-analysing the pore structure of hydrated cements, the degree of filling of the pores with water, and the mobility of water in the (partially saturated) material. The results will allow better prediction of the durability and performance of cements, e.g. in wetting/drying or freeze/thaw cycling. The study brings together two of the leading groups in the area of proton NMR – the University of Surrey, UK (P. McDonald) and the Ecole Polytechnique, France (J-P. Korb) – and will make use of the latest developments in NMR field-cycling and relaxation methods.

Organo-Aluminate Interactions
This project aims to tackle one of the most difficult problems relating to cement-superplasticisers interactions in concrete, namely, the precipitation of the superplasticiser (SP) in an inactive form (as a so-called "organo-aluminate phase") during the early stages of concrete mixing. This phenomenon, which is well known, leads to excessive SP dosage requirements in many concretes, and also in serious difficulties in predicting the right dosage when the cement or the admixture or the mixing conditions are changed. The work aims to form and characterize the organo-aluminate phases formed, the kinetics of precipitation, and its potential reversibility. It is hoped that the results will lead both to the development of better (more robust) SPs, and better prediction of cement-SP interactions in real concretes. The work is conducted jointly by SIKA (R. Flatt), ESPCI (H. Zanni, H. Van Damme,) and Lafarge (Ch. Vernet).

The partner projects of the Network range from the fundamental, such as that of the University of Bourgogne on the *effect of ions on cohesion and structure of C-S-H at nano-scale*; to the applied, such as that of the Danish technological Institute to understand the mechanisms governing the *Aesthetic appearance of concrete* over the life of a building or structure.

Above all the Nanocem network is about building a new approach to research on cementitious materials. Traditionally small, often isolated, research groups make a proposal to a national funding council where it is evaluated in competition with other proposals across the range of materials science or civil engineering. If successful, the research is carried out with the limited means of the research group on local raw materials and the results presented at conferences – often in short presentations with limited discussion. It is difficult to get long term funding for research beyond the length of one PhD study, there is considerable ignorance about work going on in other European countries and much work (especially where things did not work as planned) is never published. This all leads to considerable duplication of research effort and numerous parametric studies where the results are only applicable to the particular combination of raw materials studied.

The Nanocem approach is to put more effort into defining research projects, drawing on the collective experience of the partners to identify key gaps in knowledge, bring together previous research experience, including a critical evaluation of the techniques and experimental methods which do and do not work. The research itself is enriched by the possibility to study the same raw materials across a range of techniques. The regular review and discussion of results helps to optimise the research effort and finally productive research areas can be developed across a series of individual PhD projects.

On the scientific level the research is structured according to the processes which underlie the formation of cementitious materials by hydration:

- the formation of an *assemblage of hydrates;*
- the resulting *pore structure*;
- the *rheology* of fresh materials
- the **reactivity** of cements and SCMs;
- the relation of the micro and nano structure to durability
- and macroscopic properties.

However these must be related to the areas in which improvements can be made, particularly with respect to the *sustainability* of cementitious materials (for example through the more extensive use of SCMs); the *usability* (as seen with the recent development of self placing concrete) and the *multifunctionality* (for example, capacity for self cleaning and the elimination of pollutants achievable through the incorporation of photocatalytic materials.

Through this approach it is hoped, that in the long term, concrete will not only be the cornerstone of the construction industry for roads, bridges and buildings. It will also be a material which:

- remains clean and beautiful year after year in all climates;
- can reduce the noise level of roads:
- improve the quality of air in cities;
- can actively contribute to the comfort of our homes.
- can tell us if it is experiencing excessive stress and starting to develop microcracks

- would be able to cure itself rapidly;
- maintains structural integrity on heating and impedes the spread of fire.
- would remain fluid as long as needed and, when activated would start hardening in an instant.
- available at a modest price.

Such materials are not a wild dream. They may come into existence, provided a collective and **integrated fundamental research effort** is made on the micro and nano mechanisms operating in of concrete and other cement-based materials.

The current partners of the NANOCEM consortium are listed below. Anyone wishing to know more is invited to contact the co-ordinators at LMC / IMX / STI, /EPFL, Station Postale 12, 1015 Lausanne, Switzerland; nanocem@epfl.ch.

NANOCEM PARTNERS

Academic Contractors

1. Ecole Polytechnique Fédérale de Lausanne, Switzerland, Co-ordinators
2. Ecole Supérieure de Physique et Chimie de la Ville de Paris, France
3. Ecole Polytechnique, France
4. Université de Bourgogne, France
5. University Court of the University of Aberdeen, UK
6. Consejo Superior Investigaciones Cientificas, Spain
7. Leeds University, UK
8. Bundesanstalt für Materialforschung und –prüfung, Germany
9. Slovenian National Building and Civil Engineering Institute, Slovenia
10. Eidgenössische Materialprüfungs- und Forschungsanstalt, Switzerland
11. University of Surrey, UK
12. Technical University of Denmark, Denmark
13. Lund Institute of Technology, Sweden
14. University of Aarhus, Denmark
15. Universitat Politecnica de Catalunya, Spain
16. Danish Technological Institute, Denmark
17. Czech Technical University in Prague, Czech Republic
18. Imperial College of Science Technology and Medicine, UK
19. University of Kassel, Germany
20. Commissariat à l'Energie Atomique, France
21. University of Florence, Italy

Industrial

22. Lafarge Laboratoire Central de Recherche, France
23. Holcim Group Support Ltd, Switzerland
24. CTG SPA, Italy
25. Heidelberg Cement AG, Germany
26. Aalborg Portland, Denmark
27. Verein Deutscher Zementwerke e.V., Germany
28. Elkem ASA Materials, Norway

29. SIKA Technology AG, Switzerland
30. OXAND S.A., France
31. ATILH, France
32. Salonit Anhovo, Slovenia

FUTURE OF CONCRETE
Vision and Challenges

Arnon Bentur[1], Sidney Mindess[2] and Amnon Katz[1]
[1] Technion, Isreal
[2] University of British Columbia, Canada

1. INTRODUCTION

Prediction of the future is a difficult task, and the future of concrete is no exception. In making such predictions there are several routes one might consider. In 2002, a projection of the future was presented in a comprehensive report "Roadmap 2030: The US Concrete Industry Technology Road Map" [1], in which a wide range of future technologies and technological developments were identified, in five main categories: design and structural systems, constituent materials, concrete production, delivery and placement, and repair and rehabilitation. This report is very comprehensive in terms of the individual technologies, and the workshop participants are probably familiar with it.

In view of this available document and analysis, we decided to develop a complementary and somewhat different approach, which addresses more the trends which may drive changes, and thereafter identify, or speculate on the nature of changes that may occur. The approach taken will thus address more the trends which we think may take place and drive new technological solutions, rather than identifying individual technological developments, many of which are already outlined in Roadmap 2030 [1].

In addition, when considering drivers and the resulting changes, one needs to take into account time frames. For the present discussion we will make the distinction between medium term (next 5 to 15 years) and long term (more than 15 years). Most of the discussion will focus on the medium term where we will assume that the changes are going to be incremental. By incremental we do not necessarily mean minor changes, but rather that construction with concrete will be based on relatively massive elements, with cross sectional dimensions in the range of 0.1 to 0.3m, as we know it today.

In the long run we need perhaps to consider a radical step change which will provide the technologies with which we could produce structures with cementitious materials in which the cross section will be an order of magnitude smaller. Such a drastic change will be associated with the reduction in annual cement consumption from about 1ton per capita to about 100kg per capita. This "step jump" needs to be considered if we are to drastically reduce the environmental impact associated with the production of such huge quantities of materials (sustainability consideration), as well as improved efficiency, by moving to lightweight construction, with all of the relevant implications for manpower savings on top of resource and materials savings. This will require a material of drastically different properties than the concrete we know today. Such a development needs to be considered in a much wider context, whether concrete will maintain its position in the long run as the world's main construction material, or whether it may lose its place to an emerging new material.

2. DRIVERS FOR CHANGE

In order to better focus on the future of concrete and identify the issues which are of greatest significance for predicting changes, it might be useful to consider the role of concrete in the chain of production and use of structures and the drivers for change which may affect each of the links.

The key elements in this chain are:

- Raw materials and concrete technology (mix design and production)
- Structural design and concrete specification
- Construction with concrete
- Concrete during the service life of the structure

The main drivers for change are ever increasing requirements for sustainability and efficiency. Although these two sound as buzz-words and "motherhood" statements, they can guide us in identifying and focusing on the more critical issues where change may occur. Preliminary focusing can be established by considering the impact of these drivers on the links in the chain outlined above.

Raw materials – S*ustainability* is the prime driver here, in view of national policies and regulations and international agreements on the need to preserve the environment. These policies are clearly affecting the availability of raw materials for concrete production and their quality, in particular the bulk components, cement and aggregates. The impact of these issues will affect the future and will be considered below, in particular in the section on concrete technology.

Structural design and concrete specifications – *Efficiency* is the major driver here; integration of concrete materials design with structural design (see Section 4) can potentially be extremely advantageous. The integration should address and take into consideration all the performance characteristics required from concrete, starting from early age up to long term durability performance.

Construction with concrete – *Efficiency* is the main driver here as well, and it leads to marked improvement in the properties of fresh concrete, with the most recent one being SCC. Concrete can be produced and delivered by fully mechanized processes (ready mix, pumping, etc.) which can be extended to be automated. The drivers and the dynamic developments are quite obvious and extensively reported, especially with regards to the development of advanced dispersants and rheology modifiers. These aspects will therefore not be addressed here.

Concrete during the service life of the structure – *Sustainability* is the main driver here, and the resulting technological developments are to a large extent in the area of durability and repair. These two aspects will be treated in Section 3.2.

The expected changes in each of the areas identified above have clear-cut technological implications. However for such developments to take place, there will be a need to develop a "belt" of support, and an environment which will enable these technological advancements and innovations to occur and to be properly implemented. Perhaps the most important issues of this kind are the advancement of the performance concept and standards for concrete (Section 5), integration of materials and structural design (Section 4) and the development of skilled personnel and their training (Section 6). These will be highlighted below.

3. SUSTAINABILITY AND CONCRETE TECHNOLOGY

In the long run, one may predict that an integrated approach to sustainability will develop, based on the quantification of the environmental impact of the whole building process, by a variety of means. One such approach is based on calculation of characteristic indicators using a methodology to evaluate the environmental impact (i.e. load), as shown schematically in Figure 1. The application of such an approach was recently demonstrated, to quantify the environmental impact of various components in the construction stage of reinforced pavements, and to compare between different reinforcing technologies, steel and FRP (Figure 2) [2]. In this evaluation a procedure known as Eco-indicator 99 was used. In this calculation, the environmental impact is quantified in terms of "points" which present a normalized and weighted indicator expressing the cumulative damage in three categories: human health, eco-systems and depletion of resources. The same approach can be extended to include the whole life cycle as is shown in Table 1. When such an approach will be implemented one may expect that it will provide incentive for innovation and implementation of new technologies (see additional discussion in Section 7).

The approach shown in Figure 1 and its quantification by the Eco-indicator 99 represents the impact of sustainability from the societal point of view, and is often referred to as life cycle assessment (LCA). A more limited approach which represents the impact of sustainability on the owner of the structure is quantified in terms of life cycle cost (LCC), which is used as a tool for comparing between different design options. This approach quantifies the long term economic advantages to the owner, but is not necessarily consistent with the overall environmental benefits to society estimated by LCA methodologies.

Table 1: Comparison of environmental load (Eco-indicator 99 points) of steel and FRP reinforced pavements (after Katz [2])

Reinforcement	Environmental load, Eco-indicator 99 points			
type	Construction	Maintenance*	Disposal	Total
Steel	179000	n x 13200	6020	291000
FRP	117000	N/A	7680	124000

* Only the part that refers to steel corrosion; n = number of maintenance activities (n = 8)

It might be expected that in the near future such a comprehensive LCA driver will not have a marked impact. Yet, incremental changes will still take place in view of a range of "point" policies which are implemented such as limitations on CO_2 emissions [3]. We will discuss some of these developments in terms of two different time scales of influences on the life of the structure: the construction stage and the service stage.

3.1 Construction stage

In the construction stage the main consideration would be largely the environmental and ecological impact of the production of massive quantities of raw materials for concrete, mainly cement and aggregates. The issues and problems involved, such as the use of land for quarrying, the ecological limitations on quarrying natural sand, energy consumption in burning, CO_2 emission, etc, are well known and documented [2-4] and there is no need to describe them here.

Constraints of this kind are being turned into actual drivers for change through legislation which sets limits within which we are forced to guide the production of raw materials. An alternative approach which is developing is based on setting policies having fewer legislative limitations but which instead take the course of evaluating the environmental damage of each process, resulting in fines if it does not meet specified requirements, or providing credit points which can be turned into financial gains if it is environmentally friendly. The latter approach provides incentives for optimization based on direct financial considerations. It is becoming more common and will certainly have an impact on the concrete industry and resulting technologies.

The consequences of these limitations and policies lead most often to a search for solutions in which attempts are made to develop technologies for the production of raw materials of similar performance to the current ones, but with reduced environmental impact. In the case of Portland cement many of the potential options have already been largely utilized (e.g. alternative fuels, energy saving in the production process – preheaters, etc.). Additional developments will require modification in the clinker and cement composition, such as Belite cements, sulfoaluminate cements and high volume pozzaolan, slag & fly ash binders are examples of this approach.

Development of such modified cements with the same performance as the current ones, is fraught with difficulty, and will require intensive development effort as well as modifications in large scale production processes. The alternative outlined here, which we believe is more likely to develop in the future, is that rather than "fighting" to maintain the level of performance of the current raw materials, we will let their performance "slip", and instead achieve the performance required by adjustments to the concrete itself. We should keep in mind that our "product" is concrete, and it is its performance that we should be after. The intermediate ingredients (cement, aggregates, admixtures) are "links" in the production process but not its target. Technically and economically it may be easier to achieve the necessary performance by improving the design of the concrete itself, to compensate for lower performance raw materials, in order to meet environmental and ecological requirements and constraints.

If such an approach is to take place several changes will have to be made:

1. Specifications and standards will have to allow production of concrete based on performance specifications of the concrete itself rather than on a prescription for the concrete and strict standards for the raw materials. Such standards as are presently necessary could be relaxed, to allow a wider range of raw materials, by setting minimum values below the current ones, but maintaining the performance of the concrete they make.

2. There will be a need for expanding the current technologies that are available to provide better control of the concrete performance (e.g. admixture and filler technologies).
3. There will be a need for better quality control of the concrete production and for the involvement of highly trained technologists for the mix design and QC/QA.

All of the above imply upgrading the level of concrete production so that it becomes a much more important link in the technological chain within the construction process. This would have an impact on the production facilities as well as on the skill required from the personnel involved, i.e. the concrete plant technologists.

The changes highlighted above are not entirely new in the concept that they represent, but in order to meet the expectations addressed they will need to be expanded and upgraded to a higher level than has been considered so far, as outlined below:

Performance specifications of concrete: the notion of implementing this concept has been around for some time. A first step in this direction has taken place in the new European Standard EN 206 with respect to the use of additions to provide equivalent durability performance. This concept will need to be expended to a whole range of properties. This development is required to provide a framework for technical guidance, but we should be aware of the significance of the "legal" implications. The know-how of using low performance raw materials or concretes of extremely low cement contents, to produce quality concrete is available, and there are indications that it is being used, by-passing current regulations. Examples are manufactured sands of high fines content and concretes with extremely low cement contents, all complying with strength requirements. The performance specifications are required to pave the way for such approaches, providing room for innovation in concrete technology, and at the same time providing a framework to ensure that the full range of performances is being met, and not just a single one such as strength. Additional discussion of this issue is provided in Section 5.

Technological means to enable the production of quality concrete with lower performance raw materials/low cement contents:

The gap that will open up between the declining properties of the raw materials and the demand to maintain the concrete performance and even improve it will be bridged by a much more advanced use of chemical admixtures. The use of these admixtures in combination with advanced mix design procedures will compensate for lower performance cements and lower performance aggregates and will facilitate greater use of mineral additives as substitutes for cement.

This predicted change is not entirely new, as might be concluded from some historical analysis of the development of concrete performance over time [4,5]. Let us go back to the mid- 19th century, when Portland cement concrete as we know it in the modern sense took its first strides. At that time, and over a period of almost 100 years, up to the mid-1950's, the major advance in concrete performance (that is its strength) was achieved by enhancing the performance of the Portland cement as estimated by its strength (Figure 3). This brought us to the stage where 30 to 40 MPa concrete could readily be made; this is now more or less the "standard" concrete for many applications. Beyond the 1950's, till the present time, marked strides were made in improving concrete properties, bringing us to the age of high strength concrete; this however was largely achieved not by cement modifications, but by formulation of the concretes using dispersants and fillers (Figure 4). A key element in this progress was the development of new generations of chemical dispersants combined with enhanced technical capabilities to properly formulate concrete in the plant. The concept that we are suggesting here is consistent with this development, that is of a shift from relying on the properties of the raw materials to a leveraging of the concrete technology by admixtures and advanced mix design in the concrete plant. The trend we are discussing, however, implies using a similar "family" of techniques to compensate for a wider range of raw materials properties in order to maintain the same concrete performance, as shown schematically in Figure 5.

Quality of the concrete production plants and their personnel:
The elevation of the QC/QA in concrete plants is an ongoing process, as indicated by the increasing requirements to certify concrete plants, that has been established as mandatory in some countries, especially in Europe. This ongoing change has its impact on the level of equipment and the professionals needed for technical management and operation of the plant. The trend predicted here is consistent with this change, but will force it to move several steps ahead, especially in terms of the need for high level concrete technologists within a plant to be able to accommodate more variable and a wider range of raw materials and to master advanced design tools and admixture technology to assure the ability to meet the required performance of the concrete. This change may also have an impact on the organization of the plant, to be able to cope with a variable stream of raw materials and maintain concrete performance, and to do more advanced testing of the properties of the concrete and its ingredients. Rheological evaluation and quality control may become essential [7].

The vision outlined above suggests a drastic change in the nature of concrete production in terms of the "hardware" (equipment) and "software" (skilled personnel and know-how), and obviously, the issue of cost and economy arises. The additional cost in the sophistication of the plant and its personnel, as well as the need to apply advanced chemicals, may be offset, at least partially, by the ability to use a wider range of raw materials with lower performance than the current ones. An additional balancing component may be the financial incentives which will favor the use of more environmentally friendly materials. These capabilities of the envisioned concrete production operation and personnel could be combined with an integrated approach for design of the structure and the material to bring about the cost savings which will result from efficient overall design (see Section 4).

3.2 Sustainability and concrete in the service stage

Sustainability in the service stage is usually quantified in terms of a variety of effects related to the durability of the structure. On top of this, comprehensive life cycle analysis takes into account a variety of maintenance related issues such as heating and ventilation. Within the context of the present discussion on concrete we will address only durability and repair.

3.2.1 Durability

The durability of concrete and reinforced concrete has been studied extensively, to resolve the mechanisms involved and to develop quantitative modeling to predict the life cycle under different environmental conditions. Such quantification will no doubt have an impact on our design of concrete in the future. It should be noted however, that these models usually address the concrete from the materials point of view, quantifying effects of penetration under different driving conditions (e.g. diffusion and capillary absorption) and analyzing the chemical interactions which occur due to the penetration of various species (*e.g.* Ref. 8), and the consequences to the concrete and the reinforcing steel.

Models of this kind are inherently targeted to identify processes on the materials scale. They are not intended to consider distress that may occur on the larger scale of the structure. A noted example is cracking which depends on the nature of the concrete and the overall scheme of the structure (e.g. restraint in the case of shrinkage induced stresses). It is usually implied that this type of long term performance issues should be dealt with in the design of the structure, by the structural engineer. As a result, there is a gap between the materials technologist and the structural designer; developments and advancements in these fields are not always coordinated and harmonized with each other.

Within this framework, we may identify cracking as a critical issue, which somehow falls in between the disciplines, and will require in the future a more comprehensive and integrated treatment. The penetration of fluids and gases into concrete is considered only in terms of the permeability and diffusivity of the concrete. However, cracking, even in the acceptable range of 0.1 to 0.3 mm, will lead to increases in penetration by orders of magnitude, as seen in Figure 6, which shows data compiled from the literature. The cracking issue is back in the "headlines" due to a large extent to the gap between the materials and structures disciplines: in the drive to reduce permeability and diffusivity to enhance long term performance, the so-called high performance (low w/b ratio) concretes were developed which are very impressive in their impermeability, but on the other hand are more prone to cracking because of their increased brittleness, autogenous shrinkage (Figure 7) and thermal effects. To some extent we have replaced one problem with another, and the approach now is to take these concretes for granted (as the prime material for durable structures), and open a "new front" to combat cracking.

Cracking is one example of distress resulting from a combination of materials properties and the overall response of a structure (loading, support, restraint). The logical and efficient way to deal with problems of this kind is by integration of the concrete mix design with the structural design to optimize the performance of the concrete for the specific application, not only with respect to strength.

The approach that needs to be developed in the future will require a much more balanced treatment of penetration into concrete, by considering at the same time permeability and cracking. A balanced approach of this kind may result in re-definition of high performance so that it not based solely on the strength and permeability of the material. One could think of alternatives of having concretes in the 0.40 w/b ratio range, which inherently crack less, with additional reduction in transport characteristics of these concretes achieved by other means, as demonstrated for example in Figure 8 for capillary absorption and in Figure 7 for crack control (reduced shrinkage). To facilitate such an approach, performance based specifications will have to assume a much more prominent role (see Section 5).

The integration of the materials with the structural design, which is essential not only for durability but also for economic reasons will be highlighted in Section 4. For this approach to be successful, there are needs on the structural side of the equation as well as on the materials side. The latter is consistent with the requirements we have identified in Section 3.1, namely to be better able to tailor the concretes for a variety of properties. The professional level of the concrete technologist as well as the availability of means for such flexible tailoring (technologies and mix design expertise) will become key elements. Here too, in view of the limitations on the availability and diversity of the bulk raw materials for concrete making (cement and aggregates), a key component will be admixtures and additives which will enable such flexible tailoring. Admixture and additive technologies that can control strain softening (toughening), shrinkage, sealing and more, will become important elements.

3.2.2 Maintenance and repair

As with any other type of structure, concrete structures too require a regular program of maintenance and repair (as a young student once said to me "concrete is not forever"). We often forget how important this is. According to Li and Stang [10], in countries such as Japan and Korea the annual outlay for infrastructure maintenance will soon surpass that of new construction, and North America is not far behind. For instance, it is estimated that the annual direct cost of maintenance and replacement of deteriorated reinforced concrete bridges in the United States is in excess of $8 billion. While concrete is, basically, a durable material, it may begin eventually to deteriorate for a number of reasons:

- Poor construction practices, and in particular inadequate curing
- Poor design of the structure (e.g., no provision for drainage, inadequate cover over the steel)
- Use of substandard materials without referring to their impact on properties other than strength
- More aggressive environmental exposures than expected initially

This would suggest that any designs for new concrete structures should have built into them requirements for a regular inspection and maintenance protocol, with well-defined criteria for when repairs should be undertaken.

The economic magnitude of this problem is such that we can begin to look at more innovative (and yes, more expensive) approaches to maintenance and repair. On the materials side, it may be desirable to use materials such as polymer modified fiber reinforced concretes [11], despite their relatively high cost, as they have excellent bonding properties, high strength, and sufficient ductility to withstand the inevitable stresses that arise when two unlike materials are bonded together (differential shrinkage and creep, differential elastic response to thermal or mechanical stresses). It might also be sensible to look at non-portland cement based materials for many repairs, such as fiber reinforced plastics (FRP) which may be sprayed on, or used as wraps.

There is also room to consider new generation of cement composites, and this will be further discussed in Section 7.

Within the context of maintenance we should consider the development of "smart" concrete and "smart" structures. For instance, there have been recent reports (e.g. Ref. 12) of the development of a silicon based microelectromechanical system that could be embedded in a concrete structure to transmit data about the structure's condition over time. There are already structures which are instrumented with more conventional strain gauges and which relay information on a continuous basis regarding strains. This is a trend which should become much more common in the future.

4. INTEGRATING THE CONCRETE MATERIALS DESIGN WITH THE STRUCTURAL DESIGN

Impressive and marked leaps in efficiency in any technological product can be achieved by integration of disciplines which are relevant to the technology. Concrete structures should be no exception, and there is room to consider the potential possibilities here. A promising option within this framework is the integration of the materials and structures design, to optimize the two at the same time. It seems that there are several developments in each of these disciplines which suggest that this is feasible and could yield impressive benefits to the industry.

From the materials point of view the concrete plant can potentially produce concretes of very different compositions and properties, even in relatively small quantities, thus providing a degree of freedom for materials design which exceeds anything we are aware of in other construction materials, such as steel. The ability of having a "boutique" type operation, while keeping the advantages of scale, is based, and will be further enhanced, by improving our ability to use a wider range or raw materials for generating concretes of drastically different qualities. This is achieved by advances in admixture and additive technologies, combined with more sophisticated mix design tools, and by the mode of operation of modern concrete plants which is highly mechanized in terms of dispensing capabilities and computer controlled batching. The trends identified in Section 3 will strengthen these capabilities of the concrete production and design, not only by further advances in technology but also by the elevation of the professionalism of the personnel involved. The advancement of technological means combined with skilled technologists, able to provide mixes of various properties on demand, as well as a plant that can readily produce and batch by mechanized/automated means almost any concrete composition, will provide the flexibility and degree of freedom to the structural engineer to specify a range of properties for the concrete to enable optimization of the structure. This will, of course, require re-education of the structural engineers, so that they understand what they can actually demand of the concrete.

Traditionally, the structural design of concrete was based on inputs for mature concrete, mainly strength, modulus of elasticity, shrinkage and creep. To a large extent the requirements for durability and for early age characteristics (such as prevention of cracking and strength development) were superimposed at a later stage. Common examples, frequently cited, are the eventual specification of higher strength concrete to comply with durability requirements, whereas the structural design is based on normal strength concrete (i.e. not taking advantage of the higher strength concrete that will be eventually used for construction), or the need to apply special techniques (curing, cooling pipes, etc.) in the construction stage to prevent cracking (i.e. could perhaps be avoided if considered in the design in conjunction with the materials to be used). This kind of gap can now be bridged, based on our much better understanding of the development of concrete properties over time and the ability to model them using materials science concepts (development of properties as a function of time and environmental conditions, taking into consideration the nature of hydration reactions and microstructure development). These concepts, combined with considerations of heat and moisture transfer, have resulted in impressive strides in modeling the behavior of concrete and the development of its mechanical and physical characteristics as a function of environmental conditions.

These quantified characteristics may be incorporated into structural design models (finite element based simulations) which are comprehensive in the sense that they consider the overall structure and reinforcement, as well as the development of concrete properties from time zero, and can now be used for calculation of the behavior of the structure over its whole life span, from construction to maturity. Such comprehensive modeling will be able to take as an input the properties of the concrete to assure the proper performance of the structure at early ages as well as at later ages, and thus come up with an optimized solution which includes materials and structural consideration over the entire life span. We see now some comprehensive design tools of this kind being developed and implemented.

When fully implemented, such tools will potentially provide an incentive for the production of concretes with a large variety of properties, which will need to be tailored specifically for each project. This will in turn provide incentive for advances in concrete technology: (i) develop even better means for control of properties and (ii) skilled and professional concrete technologists in the industry to be able to comply with changing requirements for properties and to be involved in the design stage.

Examples demonstrating the nature of such development are given below, using one such emerging tool of finite element simulations (Heat and MLS modules by FEMASSE):

(a) Cracking in concrete floor/slab on grade (Figure 9) -

- The stress and strength curves for concretes of different strength grade are calculated, showing intersection of the curves (i.e. cracking) in the higher strength levels, but not in the lower one.
- Conclusion: specification for higher strength in this case leads to cracking; solution – reinforcement, or alternatively specify the higher strength and require shrinkage control as an independent variable; this can be achieved by concrete mix design, shrinkage reducing admixture or other means

(b) Cracking in bridge decks (Figure 10) –

- Doubling of the conventional reinforcement can provide the means for crack control, to reduce it from level of about 0.4mm width to about 0.2mm, which is acceptable.
- Alternative solution to reduce the crack width to about 0.2 mm and avoid the need for doubling of the steel, can be based on modifying the concrete with shrinkage reducing admixture (reduction of 50% in free shrinkage) or by adding fibers to obtain a concrete with strain softening characteristics (Figure 11b)

5. PERFORMANCE SPECIFICATIONS

In recent years, as discussed above, we have made great advances in our ability to "tailor make" concretes for a wide range of special applications: concretes with compressive strengths greater than 200 MPa, self-consolidating concretes, tough and durable fiber reinforced concretes, polymer concretes, and so on. However, in producing such concretes, and even more so for the "normal" 20 to 35 MPa concretes that make up most of our concrete production, we have relied mainly on *prescriptive specifications*. That is, at least in North America, we impose requirements on such things as maximum water/binder ratios, cement types, minimum cement contents, aggregate gradings, the type and amount of mineral admixtures and fillers, and the type and amount of chemical admixtures.

While prescriptive specifications such as these have worked reasonably well in the past, when the industry as a whole was much less sophisticated than it is now, and when the special concretes referred to above had not yet been developed, they have several major disadvantages:

- They are not, in any event, a guarantee of good, durable concrete (as evidenced by the amount of "bad" concrete that we see, and by the growth in concrete-related construction litigation).
- They tend to inhibit the most efficient use of the materials potentially available to produce "good" concrete.
- They tend to stifle innovation.

It is thus essential that we move to *performance based* specifications for concrete, particularly in light of the economic and environmental pressures alluded to earlier. This would encourage the entire concrete industry (cement producers, concrete producers, design engineers and contractors) both to be more demanding in the concrete properties that they specify, and more imaginative and innovative in their selection of materials. This would also require the industry to extend the range of materials (admixtures, fillers, polymers, fibers, aggregates from industrial wastes, and so on) that are used. Performance based specifications would thus provide a means for introducing durability issues explicitly into the design both of the material and the structure.

There are, of course, some difficulties to be overcome before we can move entirely in this direction:

- The concrete industry as a whole is not ready to make the switch from prescriptive to performance specifications. In particular, there is a lack of properly trained personnel (this is especially true for small producers) to provide the necessary technical advice and QC/QA programs. Perhaps if there was a compulsory certification requirement for all concrete producers, which included a requirement for trained personnel, this would force them either to retrain their own people, or to bring in qualified engineers/technologists.
- There would have to be some procedures established for *assigning responsibility* for adequate concrete design. Currently, this "responsibility" is rather diffusely shared amongst the design engineer, the geotechnical engineer, the cement producer, the concrete supplier, the contractor, the concrete sub-trades, and perhaps others (each of whom is ready to blame all of the others for any problems that might arise). This benefits only the lawyers. It will become necessary to designate some individual as having ultimate responsibility for the quality of the concrete – an "engineer of record".
- There would have to be developed better and quicker tests not only for the materials, but also for concrete durability. We would have to move far beyond our current reliance on the 28-day compressive strength as the sole arbiter of concrete performance. There are at the moment a large number of tests for various durability problems, but most are

inadequate, in that they take too long, give ambiguous results, or only work in very particular circumstances. There needs to be a concerted research effort to develop better tests.

The challenge for us all, then, is to devise a road map and a time line for moving to performance specifications. This step is essential if we are to use our materials efficiently and effectively.

6. EDUCATION AND TRAINING

If we are to change the ways in which we produce and use concrete in any fundamental way, the drivers for change discussed above all share one need in common – a cadre of engineers and technologists who are concrete specialists. In spite of our increasingly sophisticated research activities, (and the publication of probably more than 5000(!) papers per year on concrete), we still are unable to consistently produce high performance concrete:

- There are frequent durability problems
- There is excessive cracking and spalling
- There are too many concrete failures

In addition to the lack of concrete specialists, we also, unfortunately, lack a properly trained skilled work force. We also do not seem to have an appropriate way of transferring knowledge from the research laboratory to the field; there remains a wide gulf between what we know and what we do in practice. Certainly, the mere publication of research papers in the standard journals seems not to be effective in this regard.

Unfortunately, relatively few of the professionals who either specify concrete for various applications or who design concrete structures and pavements understand the fundamental nature of concrete. Rather, they see concrete as a "black box" whose properties can be defined completely by its compressive strength, its elastic modulus, and perhaps its Poisson's ratio. This view is all too often reinforced by the North American structural design codes, which are entirely strength-based, with only a few prescriptive rules for freeze-thaw protection, for sulfate attack, and for concrete in "severe" environments. There are typically no explicit provisions for creep and shrinkage, though of course designers are expected to take these effects into consideration for deflection calculations. The concept of *toughness* is not considered at all, though this is why we add fibers to concrete.

People who truly understand concrete behavior are in short supply. Decisions on material selection and construction practice are most often left to some combination of architects, engineers, materials suppliers and contractors, with no one apparently taking overall responsibility for the concrete. This state of affairs is a reflection on the way in which engineers who end up working with concrete are educated. They are, most commonly, civil engineers who have probably had only one or two courses on "materials of construction", of which only a part can be devoted to cement and concrete. Thus, for the other changes that we envisage to be able to happen, there is need for a new curriculum to train engineers and technologists specifically to work in the field of cementitious materials.

7. LONG TERM DEVELOPMENTS

The vision outlined above "sketches" developments which are driven by needs of sustainability and efficiency. These developments, although significant in changing the industry, might be considered as incremental, in the sense that construction with concrete will be based to a large extent on the production of relatively bulk components with a cross section of the order of 0.1m, as we know it today. These changes can be considered as part of a continuous development.

For the far future, a drastic development which might be considered dramatic and leading to a giant step change, will be feasible only if we can achieve a drastic change in the properties of concrete. Technically, it is feasible to obtain cementitious materials with high compressive and tensile strength which are ductile in nature (Figure 11). Achieving these properties with costs and production methods similar to current concrete can provide such a break-through. Whether this will happen and when, is difficult to predict. For such a change to happen, there will be a need not only to develop the new materials to be cost effective, but to advance design and construction systems which will be drastically different from the ones we currently know, to enable construction with an order of magnitude less materials. In looking into such future, we have to be open to the option that these properties might be achieved with a completely different material which will replace concrete. However, here too, there would be a need for a simultaneous change in the materials, the design concept of the structure as well as the construction practices as we currently know them.

There is room however to consider a development which will accelerate the penetration of new cementitious composites and reinforcing systems even in conventional construction. This may take place if an integrated approach to environmental impact will become mandatory. The implication for such a change on the feasibility of justifying the application of new construction technologies was demonstrated by Katz [2], for reinforced concrete pavements in which the environmental impact of steel reinforcement was compared with that of FRP reinforcement (Figure 2 and Table 1). Developments of this nature could pave the way for a range of new high performance cementitious composites such as those outlined below:

- Li and Stang [10] suggested that High Performance Fiber Reinforced Cementitious Composites (HPFRCC) might be used as the matrix in reinforced concrete structures. They argue that while HPFRCC (containing two or three percent by volume of fibers) might appear to be "prohibitively" expensive, if we factor in life-cycle costs, as well as the social and environmental costs of repairs and replacement, then this should be feasible. Indeed, a whole family of similar materials with high fiber contents and very high strengths (both tensile and compressive) are beginning to appear on the market:
- RPC was developed in France in the early 1990s [15]. The quite remarkable properties of this material (compressive strengths in excess of 600 MPa) were achieved by careful control of the concrete mixture, in particular the particle size distribution of *all* of the solid materials. Optimization of the particle size distribution leads to a mix approaching optimum density. RPC contains no coarse aggregate; indeed, the maximum aggregate size is 0.3mm! This permits production of a more homogeneous material. For purposes of ductility, up to 5% by volume of steel fibers are added to the mix.

- A commercial development of RPC is now being marketed under the name of DUCTAL®. With steel fibers, compressive strengths are of the order of 150 to 180 MPa, with flexural strengths of about 32 MPa. These strengths are reduced by about 25% when polypropylene fibers are used. A somewhat similar French material, BSI®-CERACEM concrete was used to construct the toll gate roofs for the new Millau viaduct in the south of France [16].
- Another version of this technology, also developed in France, has been patented under the name of $CEMTEC_{multiscale}$® [17]. It is characterized by much higher cement and fiber contents than DUCTAL®, though the underlying principles are the same. This material can achieve flexural strengths of about 60 MPa. It also has an extremely low permeability. For comparison, typical mix proportions for two of these materials are given in Table 2.
-

Table 2. Compositions of some commercial RPCs

Material	*DUCTAL® (kg/m³)*	*$CEMTEC_{multiscale}$® (kg/m³)*
Portland cement	*710*	*1050.1*
Silica fume	*230*	*268.1*
Crushed quartz	*210*	-
Sand	*1020*	*514.3*
Water	*140*	*180.3*
Fibers	*40 – 160[a]*	*858[b]*
Superplasticizer	*10*	*44*

[a]Either steel or polypropylene fibers (13mm x 0.20mm)
[b]A mixture of three different geometries of steel fibers

- Still another family of ultra high strength concretes was developed in Denmark in the 1980s. This material is referred to as CRC (Compact Reinforced Composite). It too is made with a very low water/binder ratio (~0.16 or less), and contains from 2 to 6% steel fibers, providing matrix strengths of 140 to 400 MPa. It differs from the materials described above in that it is also combined with closely spaced conventional steel reinforcement. It has been used mainly in precast elements such as staircases and balcony slabs [18] but has also been used in cast-in-place applications.
- There are other ultra high strength concretes that have been produced, and still others in the development stage. Their common features are a very low water/binder ratio, the use of silica fume and superplasticizers, high contents of fibers, limitations on the maximum aggregate size, and careful control of the particle size distribution. They also require very tight quality control both in their production and in their placement. Consequently, these materials are very expensive and the initial cost of construction is high. The application of such systems and the drive for the development of new ones will be dependent to a large extent on the changes in the societal values towards sustainability considerations which will be

"translated" into "down to earth" quantitative requirements to design and build structures that will meet specified life cycle criteria and/or quantifiable environmental loads.

REFERENCES

1. Strategic Development Council, Roadmap 2030: The U.S. Concrete Industry Technology Road Map, USA, 2002,

2. Katz, A., Environmental impact of steel and FRP reinforced pavements, ASCE J. of Composites for Construction, in press.

3. Gartner, E., Industrially interesting approaches to "low CO_2 cements, Cement and Concrete Research, 34, 2004, 1489-1498.

4. Aitcin, P.-C., Cements of yesterday and today, Concrete for tomorrow, Cement and Concrete Research, 30, 2000, 1349-1359.

5. Bentur, A., Cementitious materials – Nine millennia and a new century: Past, present and future, ASCE J. of Materials in Civil Engineering, 14, 2002, 2-22.

6. Bentur, A., High strength/high performance concrete for long term durability performance", Construction Specifier, 57, 2004, 66-72

7. Banfill, P.F., The Rheology of fresh cement and concrete, Concrete Plant International, April 2004, 86-100.

8. Marchand, J., Modeling the behavior of unsaturated cement systems exposed to aggressive chemical environments, Materials and Structures, 34, 2001, 195-200.

9. Attiogbe, E.K., See, H.T. and Miltenberger, M.A., Cracking potential of concrete under restrained conditions, Advances in Cement and Concrete, Proceedings from the Engineering Foundation Conference, D.A.Lange, K.L.Scrivener and J.Marchand, editors, The Engineering Foundation, 2003, 191-200.

10. Li, V.C. and Stang, H., Elevating FRC material ductility to infrastructure durability, in di Prisco, M., Felicetti, R. and Plizzari, G.A., *Fibre-Reinforced Concretes BEFIB 2004*, RILEM Proceedings PRO 39, RILEM Publications, Bagneux, France, 2004, Vol.1, 171-186.

11. Xu, H. and Mindess, S., The flexural toughness of high strength fiber reinforced concrete with styrene-butadiene latex, in Zingoni, A. *Progress in Structural Engineering, Mechanics and Computation,* Proceedings of the Second International Conference on Structural Engineering, Mechanics and Computation, Cape Town, A. A. Palbema Publishers, 2004, Chapter No. 246 (CD-ROM).

12. Binns, J., Passive devices monitor bridges, buildings, Civil Engineering, July 2004, p. 33

13. Baetens, B., Sclangen, E., van Beek, T., Roelfstra, P. and Bijen, J., Computer simulation for concrete temperature control, Concrete International, December 2002, 43-48.

14. Schlangen, E., Lemmens, T. and van Beek, T., Simulation of physical and mechanical processes in concrete floors and slabs, in Concrete Floors and Slabs, Proc. Int. Seminar, R.K.Dhir, editor, UK, 2002, 45-56.

15. Richard, P. and Cheyrezy, M. H., Reactive powder concrete with high ductility and 200-800 MPa compressive strength, in Mehta, P. K. *Concrete technology past, present and future,* SP-144, Farmington Hills, MI, American Concrete Institute, 1994, 507-518.

16. Thibaux, T., Hajar, Z, Simon, A. and Chanut, S., Construction of an ulta-high-performance fibre-reinforced concrete thin-shell structure over the Millau viaduct toll gates, in di Prisco M, Felcetti R and Plizzari G A *Fibre-reinforced concretes BEFIB 2004,* Vol. 2, Bagneux, France, RILEM Publications, 2004, 1183-1192.

17. Parent, E. and Rossi, P., A new multi-scale cement composite for civil engineering and building construction fields, in *Advances in concrete through science and engineering,* Bagneux, France, RILEM Publications, CD-ROM Paper No. 14, Hybrid-Fiber Session, 2004.

18. Aaarup, B., CRC – a special fibre reinforced high performance concrete, in *Advances in concrete through science and engineering,* Bagneux, France, RILEM Publications, CD-ROM Paper 13, Hybrid-Fiber Session, 2004.

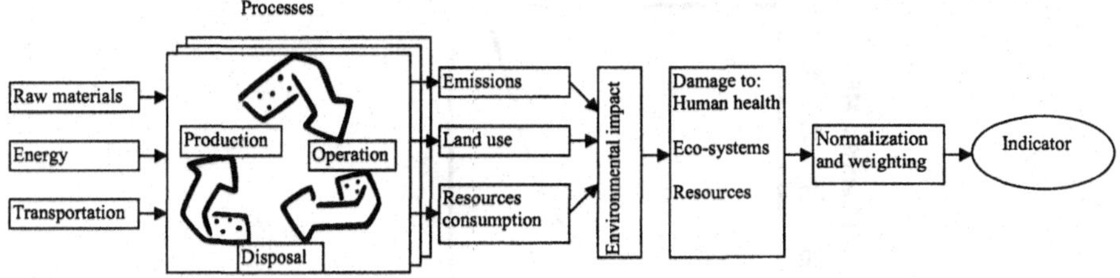

Figure 1: Flow of environmental assessment of processes (after Katz [2])

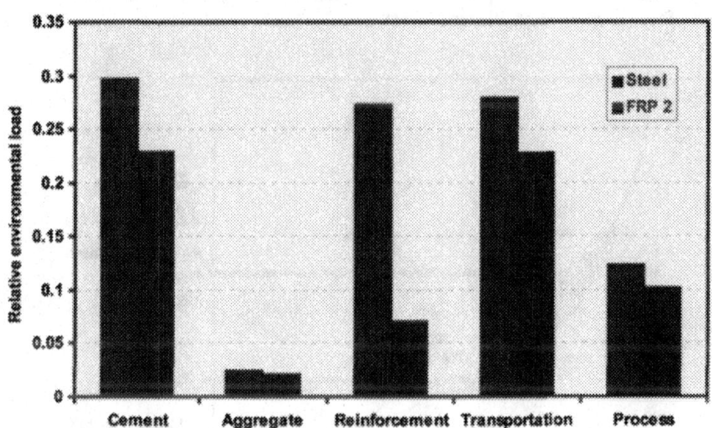

Figure 2: Comparison of environmental load of pavement construction with steel and FRP reinforcement, using Eco-indicator 99 method, normalized to load of steel reinforced pavement (adopted from Katz [2])

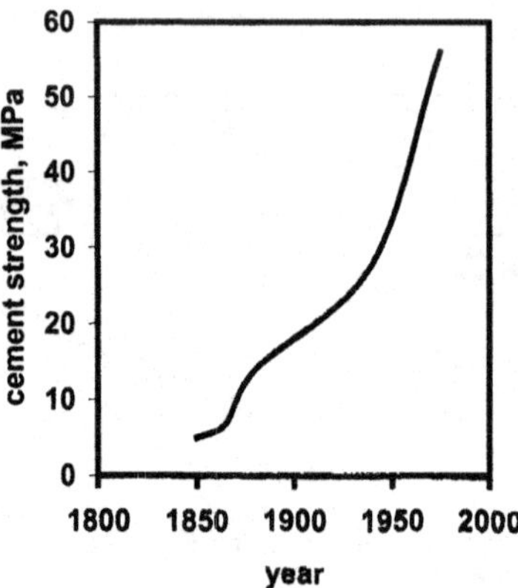

Figure 3: Development over time of performance of Portland cement as characterized by strength of 1:3 mortar cubes (adopted from Blezard, in ref. 5)

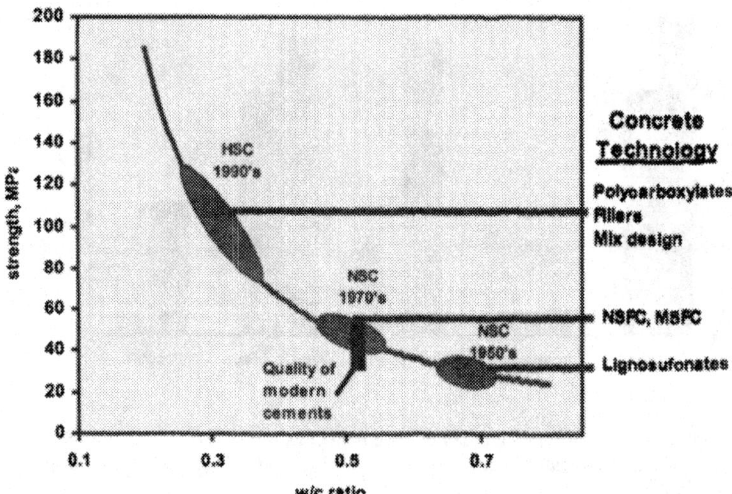

Figure 4: Development in concrete strength over time in conjunction with newer generations of dispersing admixture (adopted from ref. 6)

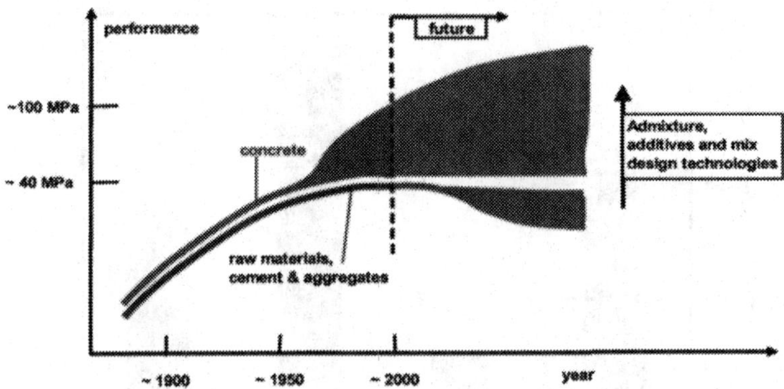

Figure 5: Schematic description of the development of the qualities of concretes and cements until recent times, and prediction to the future.

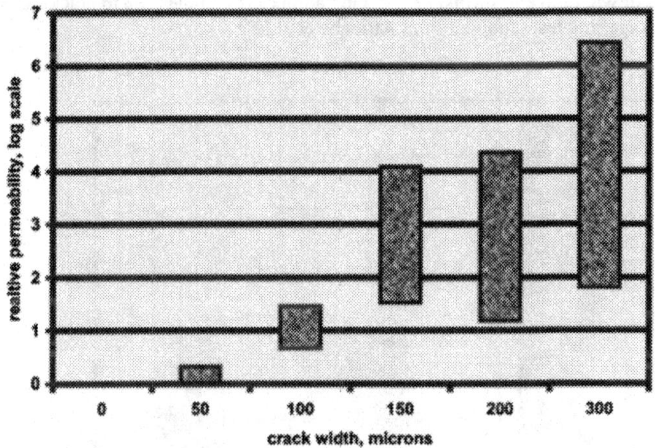

Figure 6: Range of relations between crack width and permeability compiled from data in the literature (adopted from Ref. 6)

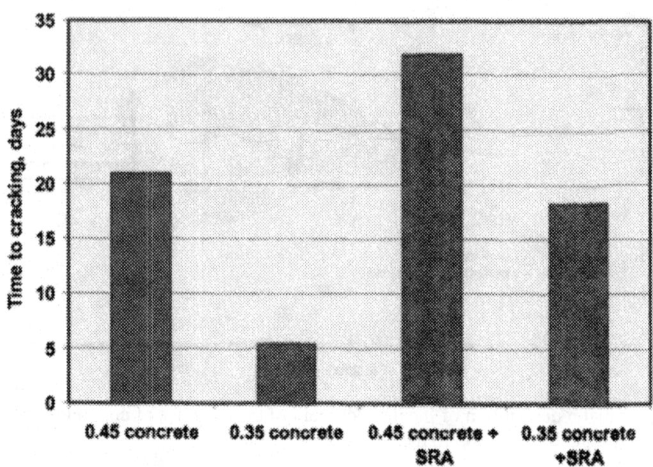

Figure 7: Effect of w/c ratio and shrinkage reducing admixtures on the time to cracking in a restrained shrinkage ring test (adopted from Attiogbe et al [9])

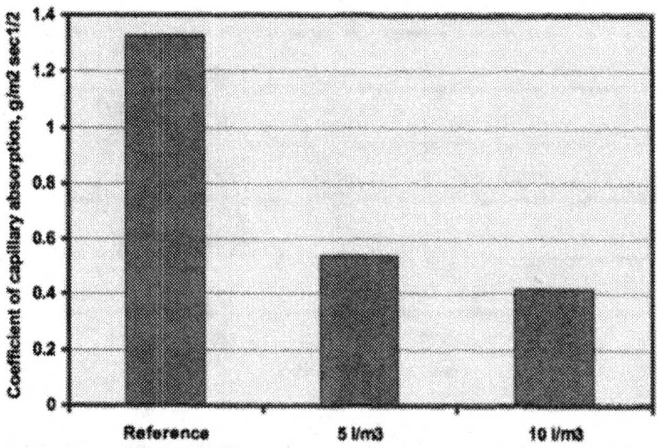

Figure 8: Effect of damp proofing admixture content on the reduction in the absorption of 0.40 w/c ratio concretes (adopted from data of Berke and Li, in ref. 6)

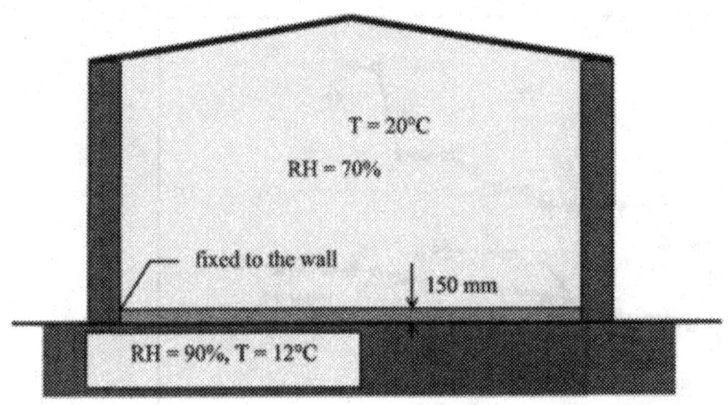

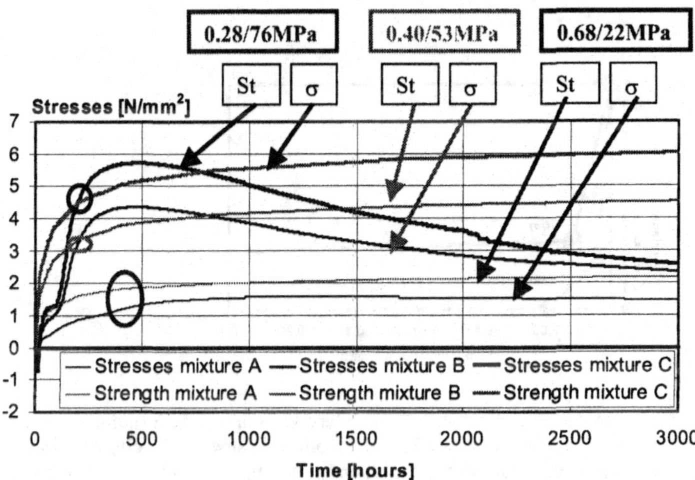

Figure 9: Simulation to calculate the stresses (σ) and strength (St) development in a slab (a) showing intersection and cracking in the higher strength-low w/c concrete and no-intersection at the lower strength concrete (b) (after Schlangen et al. [14])

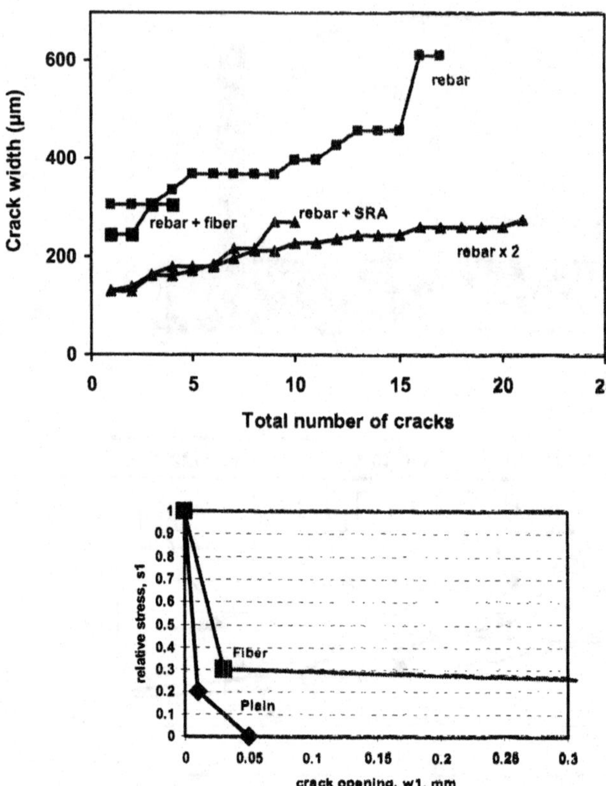

Figure 10: Simulation to calculate the development of cracks in bridge deck (number and width of cracks along the deck – a), showing the effect of doubling the steel reinforcing or using concretes of enhanced properties (shrinkage reduction by 50% using shrinkage reducing admixture or fiber reinforcement to obtain strain softening shown in b). Adopted from analysis of Li, W.R.Grace

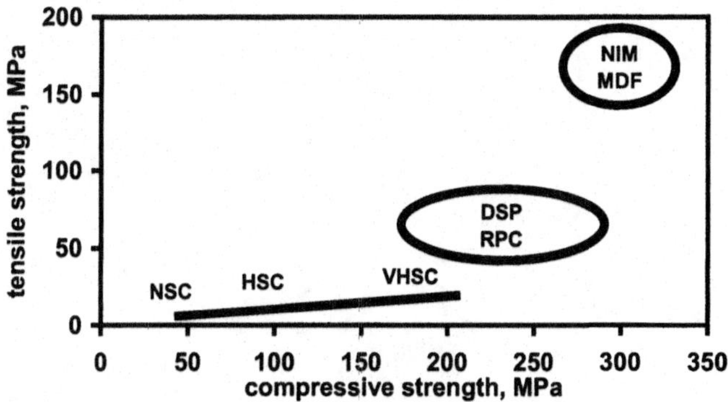

Figure 11: Relations between tensile and compressive strength of conventional concretes and newly developed high performance composites systems (after Bentur [5])

SDC, AN INDUSTRY TOOL TO IMPROVE THE EFFICIENCY OF THE CONCRETE INDUSTRY

Peter Emmons
Co-Chairman SDC
CEO Structural Group
Maryland, USA

The SDC (Strategic Development Council) is an eight-year-old organization developed by ACI to facilitate the accelerated use of new technology to improve the overall efficiency of the concrete industry. Two years ago the SDC began the development of a program called ATA (Accelerating Technology Acceptance); like the name says it is all about finding ways to move new and existing technology into the mainstream faster. A key component of the ATA program is identifying the top technology issues, which have the greatest impact to improving the efficiency of our concrete industry. By identifying the most important technology issues, the limited resources available can be better allocated to the top issues. The SDC is in the process of identifying the top technology issues by

SDC VISION STATEMENT

Bring together the concrete industry, along with government, academia and customers, to focus on collaborative problem-solving in meaningful technology advancement. The result shall be the expeditious movement of meaningful innovations through the standardization process into commercial use.

canvassing industry participants about what they each feel are important issues. The identified items are then prioritized based upon a number of factors including:

Financial Impact

A significant factor in rating/ranking the importance of a technology/issue to the concrete industry is its economic impact to the industry. Provide a rational description and estimated financial impact statement. Develop an estimated return on investment (ROI).

Scope

The technology/issue should be defined so that it is concentrated in scope and of a feasible size that the concrete industry can assemble and focus resources to realistically deal with the technology/issue. Identify the sphere of applicability. Provide a scope of the technology/issue.

SDC FUNCTIONS

- Innovating
  - Research Consortia
- Visioning
  - Vision & Roadmap 2030- Concrete Industry Plan
  - ATA Program - Accelerating Technology Acceptance
  - Vision & Roadmap 2020 –Concrete Repair Industry Plan
- Facilitating
  - Top 20 Technology Issues
    - P2P
    - Floor Moisture
    - Building Code Simplification
- Enabling
  - Technology Showcases (@SDC meetings)
  - Showcase Projects
  - Customer Forums
- Coalescing "Defragmenting the Industry"
  - Networking, Visioning Together

Industry Effect

Technologies/issues which affect a larger and broader base can rally support and resources from many sources. Demonstrate the breadth of impact of the technology/issue – a segment of the concrete industry, the whole concrete industry, or entities beyond the concrete industry including owners, manufacturers, alternate material suppliers.

Immediacy/Urgency

Technologies/issues which promise more immediate return should be addressed first. Describe the risk, liability, and lost opportunity costs in delayed technology acceptance/ problem resolution.

SDC MISSION STATEMENT

The SDC shall facilitate advancement of concrete technology by:

- Providing a forum for visioning and prioritization of key challenges facing the concrete industry
- Assisting ACI in timely adoption of innovative technologies, problem solutions, and best practices
- Providing a framework for consortia-directed development programs leading to or supporting improved commercialization
- Providing appraisal services when needed

Sustainability Factor – Environmental / "Green"

Does the technology/issue mitigate an existing environmental problem or promote sustainability of a building/installation? Identify positive and negative impacts to the environment. Identify the life cycle cost savings.

Sustainability Factor – Concrete Industry Competitiveness

Does the technology/issue enhance the competitiveness of the concrete industry? Is the efficiency of the industry improved? Is the resulting product/installation more desirable? Provide narrative of how the technology/issue resolution contributes to sustaining the concrete industry.

Quality Issues

Does the resulting product/installation affected by the technology/issues meet/exceed the needs and requirements of owner/user? Is the functionality of the product/installation improved relative to the savings/additional cost of implementing/using the technology/resolving the issue? Define the value added or cost savings.

Other

Describe other qualitative characteristics of the technology/issue that are of immediate and significant importance to members of the concrete industry, the concrete industry overall, the construction industry, and to the owner/user.

During the next few months the SDC will announce the top issues along with a strategic plan for each issue. It is the objective that after identification and planning, resources will be identified and leadership applied to facilitate the acceleration of technology acceptance.

THE P2P INITIATIVE

Jeff O'Leary
Director, Technical Services
Florida Rock Industries, Inc., USA

Lionel Lemay
Senior Director of Applied Engineering
National Ready Mixed Concrete Association, USA

1. WHAT P2P IS

P2P is an acronym for Prescription to Performance specifications for ready-mixed concrete. Current specifications tend to be prescriptive in nature and limit even some of the most basic advancements in concrete technology. This causes concrete construction to be less efficient and less cost effective. Understanding that the concrete construction industry will better be served through the use of advanced technologies, the ready mixed concrete industry has initiated an effort to develop an alternative, performance based specifying method.

The P2P initiative is a product of the National Ready Mixed Concrete Association's (NRMCA) Research, Engineering & Standards committee. A steering committee was formed in October of 2002 and immediately established task groups to focus on specific segments of our industry. The task groups include:

- Engineers/Architects
- Contractors
- Regulatory
- Associations
- Communications
- Materials

The steering committee and task groups are working together with these critical stakeholders to insure the entire concrete construction industry, not just producers, is involved with this initiative.

2. WHAT P2P ISN'T

The P2P initiative is not an effort to force a change which would not be beneficial to all parties nor is it an attempt to eliminate prescriptive specifications completely. Prescriptive specifications certainly may continue to have their place in construction.

The initiative is also not an attempt to remove the specifier's input from the mixture proportioning process nor hide vital information from the specifier. The specifier will always be involved with the process; he will simply specify his design criteria in lieu of using other, more vague criteria to achieve the performance he needs.

3. OBJECTIVES

The intent is to have an acceptable, alternative specification, which improves the quality and competitiveness of concrete construction. With alternative specifying methods, the marketplace will determine the effectiveness of performance based specifications.

Objectives of the initiative include:

- Establish a credible alternative to current prescriptive specifications.
- Utilize the expertise of all stakeholders to improve quality and reliability of concrete construction.
- Better establish roles and responsibilities based on expertise.
- Solidify the credibility of the ready-mixed concrete industry.
- Allow for innovation and new technologies with concrete mixtures and construction means & methods.

4. PRESCRIPTIVE SPECIFICATIONS

Prescriptive specifications limit the use of beneficial raw materials, control concrete mixture proportions and restrict construction methods.

Inherent problems with prescriptive specifications include:

- They do not always address the intended performance of the concrete. If low shrinkage is needed, a low w/c ratio specification will not replace a maximum shrinkage requirement.
- They typically do not provide any incentive for the producer to control the consistency of the concrete. Meeting prescriptive specifications typically requires very little, if any, quality control of the concrete properties.
- Producers are relegated to simply batching the proportions specified. The producer is given limits by which to batch the concrete and is rarely allowed to make adjustments, even though his raw materials change.
- The producer is usually held responsible for problems or defects, even though he lacks the freedom to make changes. If weather conditions change, but the concrete mixture is not adjusted, problems will occur with setting time, strength gain, and the like.

Most specifications use prescriptive means to achieve implied performance criteria. The specifier uses what appear to be simple tools to control performance criteria of the concrete, such as a maximum water/cementitious ratio, minimum cementitious content, maximum slump, etc. However, these requirements many times conflict with the intended performance, such as a minimum cement content requirement, which may actually increase cracking.

Currently, most projects, which utilize prescriptive specifications, are awarded to the lowest bidder. This methodology inherently results in the bid being awarded to the producer with the least investment in research, testing and quality control.

With low permeability being one of the most critical characteristics of concrete, specifiers typically rely on a low water/cement ratio to control this property. However, this can result in a significant increase in the cement content, resulting in greater heat of hydration, increased shrinkage and resultant cracking, and higher mixture cost. The use of supplementary cementitious materials, such as fly ash, ground slag, silica fume and the like, can provide a greater reduction in the permeability of the concrete, as shown in Fig. 1 below.

Permeability vs w/cm

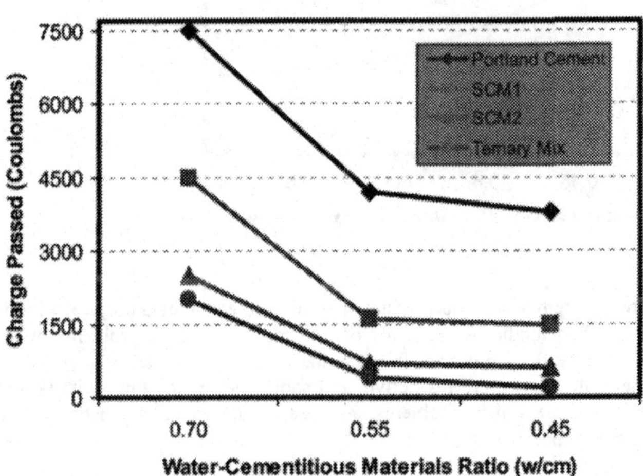

The water/cementitious ratio also is utilized to control the strength of the concrete. However, strength levels above required compressive strength usually indicate an excess of cementitious material and can result in the same problems previously mentioned. Figure 2 below shows how various mixtures with the same w/cm ratio can result in significantly different strength levels.

Strength vs w/cm

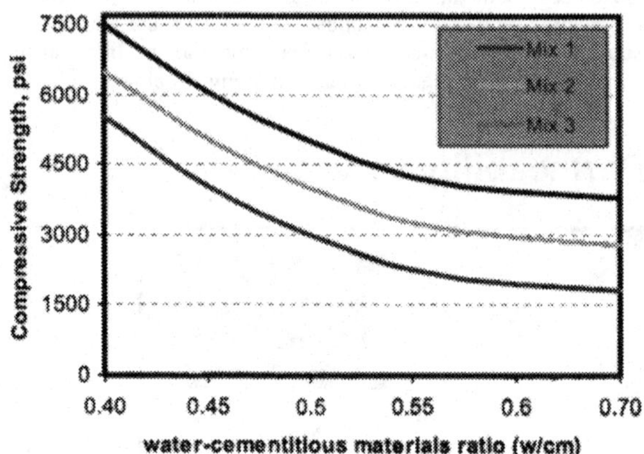

The current prescriptive system schematically shown below can also create a cycle, which puts increasingly tighter and less applicable controls on the project. When problems arise, the specifier may feel the contractor or producer failed in their duties. In an effort to prevent such problems, the specifier will insert more restrictive, and sometimes conflicting, criteria into the specifications. This then creates more problems, additional responses and subsequently more restrictions are added to the specifications.

Current System

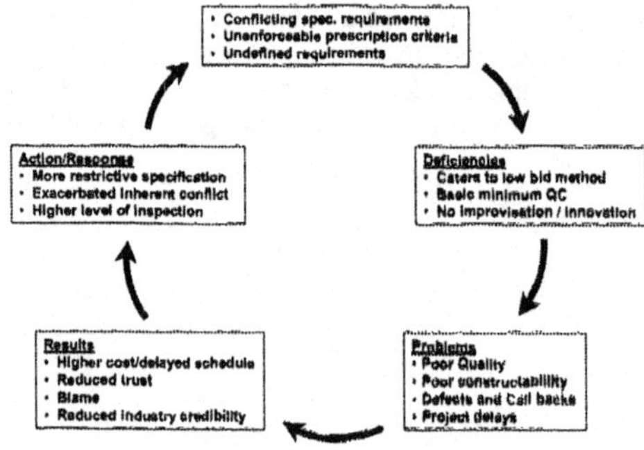

5. PERFORMANCE SPECIFICATIONS

Performance specifications focus on the functional or performance requirements of the concrete. The acceptance criteria and methods of test are spelled out such that there is no confusion and the requirements are enforceable. The specifications allow the contractor to work with the producer to develop mixtures, which meet the needs of the contractor, typically placing and consolidation criteria, and of the designer, typically strength and durability criteria. Advancements in construction technology and innovative products can be utilized with performance specifications, providing benefit to the entire construction team.

The producer will be required to prove his capability to meet the requirements of the specifications. This includes technical personnel who have been certified to proportion mixtures for performance specifications, as well as significant operational and quality control criteria of the company and its facilities.

In short, performance specifications address what the specifier needs from the hardened product. If the specifier desires minimal shrinkage, greater durability or low permeability, he specifies such. He will rely on the expertise of the producer to provide what he needs.

Restrictions put on the producer regarding mixture adjustments, will result in concrete with variable performance. Because the raw materials and environmental conditions are always changing, the concrete mixture ingredients or proportions must be adjusted to achieve consistent performance. Performance specifications allow the producer to make the adjustments which are necessary on virtually every project:

Variable Raw Materials/Environment + Fixed Proportions

= Variable Concrete Performance

Variable Raw Materials/Environment + Adjusted Proportions

= Consistent Concrete Performance

6. THE BENEFITS OF P2P

The project team must realize a benefit for performance specifications to be effective.

The owner receives a structure which is likely to have improved performance, longer service life, reduced construction time and lower cost.

The specifier can focus more on performance rather than mixture composition. The specifier will also benefit from a more efficient submittal review process, improved quality of the final product and reduced conflicts with the contractor.

The contractor will have his constructability requirements addressed with predictable performance. He will also better be able to utilize innovative construction techniques and maintain better control over the construction schedule.

The concrete producer will have much clearer specifications, reducing post-bid conflict, and would be allowed to make adjustments to the mixtures to insure the requirements of the contractor and designer are met.

7. How It Will Work

Exact method, criteria or qualifications have yet to be finalized. However, there are certainly some basic requirements of the industry in order for performance specifications to be successful.

It is anticipated that codes and standards will include both the current prescriptive requirements as well as the performance requirements as an option. The designer will have the option to select the prescription or performance option for specifying concrete. The contractor would work with the producer to develop concrete mixtures and construction methods that meet both the needs of the designer (hardened properties) as well as constructability needs (plastic properties). Other requirements inherent to performance specifications would include requirements for prequalification of the producer and acceptance criteria.

All projects will not require the same detail of specification, performance criteria, producer qualifications, submittals, etc. A small residential or light commercial project will have fewer criteria than a major, industrial project. The Quality Management system will match the magnitude of the project, in lieu of one system for all projects.

The submittal review process by the contractor and designer will be greatly reduced, yet be much more relevant than todays' practice of excessive and many times meaningless paperwork. The exact materials or proportions of the mixtures will likely not be included with the submittal but rather include a certification that the mixtures will meet the specifications and any prequalification test data.

Contractors will work with a producer to develop mixtures which meet his placement needs as well as the requirements of the specifications. The producer will have a Quality Management system in place, which has been previously approved and/or certified by a national entity. The plan will outline personnel qualifications/certifications, equipment requirements and procedures for quality assurance of raw materials as well as the quality control of the concrete mixtures.

The producer will develop mixtures and prepare submittals based on the specification requirements as well as the contractor's needs. The submittals will be prepared only by qualified personnel being certified through a national program. In order for the designer to have confidence in the mixtures the certification of personnel proportioning the mixtures is critical. While designers realize that many producers are much more knowledgeable in mixture proportioning the designer needs to be comfortable with formally recognizing the producer as the entity responsible for the mixtures.

8. CHALLENGES FACING P2P

Education of all within the industry is likely the single biggest challenge. While trust is an important issue it cannot be developed without everyone having a full understanding of how performance specifications work and what the benefits are.

Initially, many specifiers may not recognize a need for change nor desire one. Performance specifications may not work for every specifier or project but will certainly fit the scope of many projects.

Codes and standards must be addressed to assist the designers with specifying performance specifications. As we all know, this process can be relatively slow. For those specifiers who are not currently comfortable with performance specifications, a guide performance specification will be vital.

While many acceptance criteria and test methods are currently available, there is likely a need to improve upon some of these. New test methods may need to be developed, existing methods may need improvement and the relationship between some test methods and actual performance may need to be better understood. Test methods must be relatively easy to perform, cost effective, provide results in a timely fashion and practical for jobsite use.

9. ACTIVITIES UNDER-WAY

The NRMCA P2P committee has undertaken many activities, some of which have been realized. Communication with key stakeholders is critical with this initiative. Changes in the industry cannot be done without the cooperative effort of all parties. The committee has established relationships with specifiers/designers, contractors, codes/standards organizations, trade associations and producers to insure both their input with this process and their understanding of performance specifications.

Informative programs focusing on the general concept of performance specifications are currently available. More specific educational programs are developed as details of performance specifications become solidified.

Current codes and standards are being studied for changes to accommodate performance specifications. As a quality management system, acceptance criteria and test methods are finalized, the committee will formally address the codes and standards.

A guide specification is being developed to provide assistance to designers who would like to utilize performance specifications but are not familiar with the concept.

A producer quality management system is being developed. There exists today, most of the requirements of a QM system, which include training and certification of personnel for mixture proportioning, training and certification of personnel for testing procedures, and inspection and certification of concrete production facilities and equipment.

A review of current test methods and recommendations for additional research is currently under way. Changes or additions to existing methods as well as the development of new methods will require acceptance by the codes and standards.

Case studies are also being collected to evaluate the effectiveness of performance specifications.

Progress and information on the P2P Initiative will continue to be disseminated through the NRMCA website at www.nrmca.org/P2P.

10. CONCLUSIONS

Since the days of the designer detailing the number of bags of cement and units of aggregate to put into the mixer, the concrete production industry has come a long way. Concrete production facilities are fully automated with computer systems all but eliminating human error. Most producers also have complete quality control programs in place with dedicated personnel performing duties which include product development, mixture proportioning, QA, QC, and troubleshooting.

The industry has gained tremendous knowledge regarding raw materials, concrete performance and construction techniques. Specifiers have also become more specialized, concerning themselves less with concrete materials and more on the design of structures. The concrete producer has become more of an expert regarding concrete mixtures than most designers.

The producer typically better understands what it takes to achieve characteristics such as low shrinkage, high strength or low permeability. Not only does he have an exceptional understanding of his concrete, but he is also very familiar with raw materials and how they affect the concrete.

NEEDS FOR CONCRETE TESTS AND STANDARDS

R. Doug Hooton
University of Toronto, Toronto, Canada

ABSTRACT
Some of the concerns related to adoption of performance specifications are related to the lack of adequate test methods relating to performance for each of the relevant performance properties of concrete. In this contribution, these issues are discussed with respect to the current situation and new or needed developments in test methods. To illustrate this, the examples of permeability and sulfate resistance are used.

1. INTRODUCTION

There is a current trend away from prescriptive towards performance specifications in North America and around the world. Prescriptive specifications have been developed by experience and often are conservative. They often inhibit innovation since new materials and methods do not fit into the prescriptive mold. However, adoption of true performance-based specifications presupposes that we have a clear understanding of all the performance issues that can affect concrete. It also assumes that there are appropriate performance test methods in place to evaluate all of the performance issues for: concrete materials, fresh concrete, hardened concrete, and durability. It also assumes that performance can either be measured in time to affect the outcome, and/or can be used to prequalify concrete mixtures.

2. CURRENT SITUATION

Fresh Concrete Tests

Workability. The slump test, while a measure of consistency rather than true workability, is still the most widely used test in the world. It has limitations in the low slump range, where tests such as the Vebe test are more appropriate. However, that is more of an issue for dry cast, precast concrete operations. For concretes at high slumps, the advent of Self-Consolidating Concrete (SCC) has resulted in the slump flow test (the same procedure except that the diameter of the slumped concrete is measured along with the presence of any halo due to bleeding). Limits on the slump flow test and other flow tests for SCC (L-box and J-ring) have been adopted in CSA A23.1 (2004) and ASTM C09.47 is working to adopt standards for these tests. While there has been much recent work on rheology of concrete, it is unlikely that this will have an impact on testing of fresh concrete in the near future.

Air Entrainment. Currently the pressure method and volumetric method are used to determine total air content on fresh concrete. In addition, density measurements on fresh concrete can also provide some measure of uniformity of air contents. Also of concern is getting a measure of the air void distribution, but until recently there has not been a test method. There has been work by several DOT's, especially Kansas DOT, and the FHWA on the use and adoption of the Danish fresh air void analyzer (AVA) test. In this test, a sample of concrete is injected into the bottom of a column filled with a glycerin mixture and stirred. The entrained air bubble rise up in the column, at a rate dependant on their size (small ones rise slower). The buoyant mass of all the air bubbles are measured with time as they are trapped under a balance at the top of the column. This provides an air void size distribution as well as total air content. There is increasing interest in this test, as a field acceptance test, but it has not yet been standardized.

Hardened Concrete Tests

Strength tests, based on site cast cylinders, typically at 28 days of age are the most common concrete performance acceptance tests. However, the results are received 28 days too late to prevent low-strength concrete from being placed. Lower than normal densities of cylinders taken after mold removal can be used as an early indicator of low-strength cylinders if the problem is due to faulty compaction or high air contents. Pull off tests as per ASTM C 900 can be used to determine safe form or shoring removal times, which are the major issues for the contractor, but this test is not commonly used. Maturity methods, such as ASTM C 1084, can also provide useful contractor information, if thermocouples and a data logger are used to follow the temperature history of the concrete element in place. The development of strength in structural elements can then be estimated once the strength-maturity relationship is established for that mixture.

Since Duff Abrams published his famous paper on W/C in 1919, concrete has been designed to obtain specified strength. Specifications are most often dominated by strength requirements, and durability is often inadequately seen as simply being obtained together with strength. When durability requirements are also specified (e.g. using RCPT limits, or max. w/cm limits), they often necessitate the use of better quality concretes than are required for the strength part of the specification — i.e. this can create incompatible specification requirements. As a result, some contractors only order (and some producers only supply) on the basis of the lowest common denominator---- the specified strength!

Durability Tests

Most deterioration processes involve two stages. Initially, aggressive fluids (water, ionic solutions with dissolved salts, gases) need to penetrate or be transported through the capillary pore structure of the concrete to reaction sites (*e.g.*, chlorides penetrating to reinforcement, sulfates penetrating to reactive aluminates) prior to the actual chemical or physical deterioration reactions. A standard acceptance test or tests to measure transport rates, or a related index test is fundamental to the development of performance-based durability specifications.

Test methods related to measurement of various durability properties exist in various standards (e.g. ASTM, AASHTO, Corps of Engineers (CRD), individual DOT's) in North America and abroad. Limits based on some of these test methods are specified in ACI, BOCA, CSA and individual DOT specifications, amongst many others. It was reported by Roumain (2002) that in the US alone there are over 2000 specifications for concrete. Each of these specifications employs different test methods and different test limits.

Another issue is that tests do not exist for all of the relevant durability or performance concerns. As well, existing tests are not always rapid, accurate, or repeatable, nor do they necessarily adequately represent any or all of the exposure conditions in-situ. The lack of adequate performance-related test methods for concrete is one of the main factors that inhibits the move from prescriptive to performance specifications. A couple of examples related to specific durability issues are used to illustrate need for relevant test methods.

De-Icer Salt Scaling. The current ASTM C 672 scaling test does not relate very well to field performance. It unduly penalizes fly ash and slag systems which do not have enough maturity at the start of the test. The Quebec BNQ test appears to better relate to field performance (A Laval U., U. of Sherbrooke, U. of Toronto, and NRCan consortium project is currently addressing this). De-icer salt scaling issues are also becoming more complex due to the introduction of alternate de-icer salts. $MgCl_2$ is now being used as de-icer salt or pre-wetting brine. MgCl2 is known to attack both the C-S-H and CH matrix to form weak M-S-H and brucite. When tested by ASTM C 672, $MgCl_2$ does not cause much scaling, but there are indicators that there are problems in the field. Research on this is currently being done for 7 DOT's and FHWA at Michigan Technical University and University of Toronto to see if there is a scaling concern.

Sulfate Resistance. Much of the early research on sulfate resistance was undertaken in the 1920-1930's. Part of this focused on the resistance of the cement itself. In 1919, Thorbergur Thorvaldson, at the University of Saskatchewan, initiated studies and in 1927 reported that C_3A was responsible for the deterioration of cements exposed to sulfate solutions, and later that high iron cements were more resistant (In 1928, Hansen, Brownmiller, and Bogue identified this phase as C_4AF). The Canada Cement Co., who had funded the research, then patented the first sulfate resistant cement, sold as Kalicrete, in 1933.

Around the same time (1921-1928), R. Wilson and A. Cleve at the PCA were studying the influence of water to cement ratio on concretes semi-immersed in sulfate soils and water in Montrose, Colorado and South Dakota. They found that regardless of the cement used, any concrete with w/c > 0.45 was deteriorated after 7 years exposure. This is shown in Figure 1.

This was followed up by an extensive 40-year study by the USBR. The benefit of staying below a w/c of 0.45 was confirmed, as summarized by Monteiro and Kurtis (2003) and as shown in Figure 2.

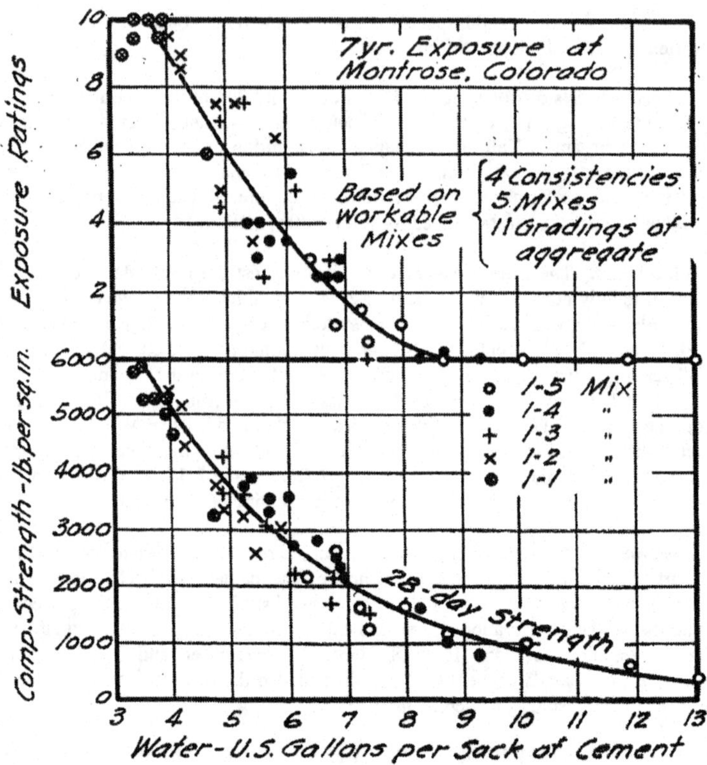

Fig. 3—Water-Cement Ratio Strength and Rating Curves.

Figure 1: Results of Outdoor sulfate exposure in Montrose Colorado (Figure 3 from Wilson and Cleve, 1928) Note: 4 gal./sack = 0.36 W/C, 6 gal./sack = 0.55 W/C, 8 gal./sack = 0.73 W/C

The w/c limits in ACI 318 are a result of this work and show that the prescriptive w/c limit is important to obtain low permeability for the prevention of sulfate attack, regardless of cement type. It has been suggested that the w/c limits be replaced by a permeability index test, such as ASTM C 1202.

In addition, the current ASTM tests only deal with ettringite sulfate attack. There are no tests for thaumasite sulfate attack. There are no tests for sulfate crystallization, but the ACI w/cm limits, help prevent an open (continuous) pore structure which allows wick action and damage due to salt crystallization.

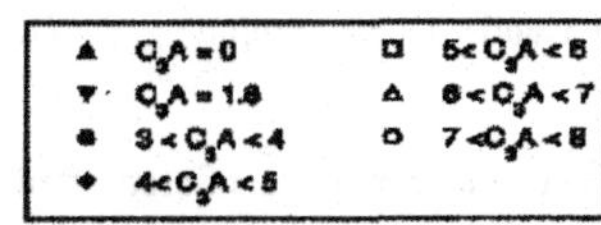

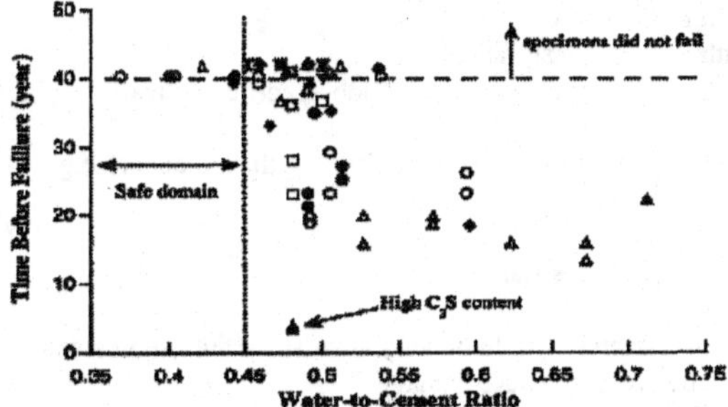

Figure 2: Time to Failure of USBR Concretes Exposed to Sulfates as a Function of W/C (Monteiro and Kurtis, 2003)

To develop relevant performance tests, any test method should result from an understanding of materials science. The mechanisms of deterioration and how the chemical and physical processes result in changing concrete properties and performance under different boundary conditions must be understood.

An example of this thinking is illustrated using the ASTM C 1012 test for sulfate resistance. It was developed to test blended cements and SCM's on a fair basis. The mortar mixtures are allowed to react and develop 20MPa strength (and some degree of impermeability) before being exposed to sulfate solutions. Sodium sulfate is used since it reacts mainly to form ettringite. Expansion is measured, since ettringite formation results in expansion. But because of these things the test is slow, since sulfates have to penetrate, then react and expand.

However, this only tests the chemical resistance of the paste fraction of the concrete. Resistance to external sulfate solutions involves the following two phases:

1. Sulfates must first penetrate the pore system of concrete.
   - Therefore, both ACI and CSA put maximum limits on W/CM as an indirect way of reducing penetrability of sulfates by provision of good quality concrete.
2. Sulfates (and cations) then react with the matrix to form alteration products which either cause swelling/cracking or loss of strength/integrity.
   -- Therefore, ACI/CSA put limits on C3A of Portland and test by ASTM C 452. Combinations of Portland cements and SCM's as well as blended cements are tested using ASTM C 1012.
   - ACI limits maximum C 1012 expansions based on severity of sulfate exposure.

However, current standard tests, including ASTM C 1012, do not address all the issues that affect performance:

- They don't account for effects of associated cations (ie. the type of sulfate)
- They don't deal with boundary conditions
- They don't deal with temperature
- They don't deal with thaumasite sulfate attack
- There is no way to test all variables, except on job-specific applications (but tests are too slow).
- But it is possible to model some of these effects with geo-codes, e.g., PHREEQE or STADIUM

Why not just test concrete for sulfate resistance?

- A performance test for concrete would be very slow since the sulfates would first have to diffuse inwards and then react to cause damage.
- Because of the different causes of sulfate damage, prisms would have to be both immersed and partly immersed, and likely at cool as well as at room temperature with different sulfate salts.
- This would be slow and expensive relative to the current situation.

Penetration Resistance Tests

We either need limits on specific transport properties (e.g. diffusion in ASTM C 1556 or sorption in ASTM C 1585) or vapor transport (ASTM E 96) or for expedience, we have to adopt a rapid permeability-index test, such as ASTM C 1202 or the Rapid Migration Test, Nordtest NT492 (AASHTO TP64). The ASTM C1202 test has become widely used for this purpose.

But while these tests can be used for job prequalification, are they suitable for quality control during construction? Various parties to construction have different interests in tests.

Are tests to be used for prequalification?
Are tests to be used for QA/QC?
Are tests to be used as input for service life models?

The new CSA A23.1-04 concrete standard has introduced ASTM C 1202 limits for prequalification of concrete mixtures to meet a) C-1 exposure conditions (concrete exposed to freezing in a saturated condition with de-icer salts, 35 MPa, air-entrained, 0.40 w/cm max.) of 1500 coulombs at 56 days, and b) C-XL exposure (like C-1 but where extended service life is required, 50MPa, air-entrained, 0.37 w/cm max.) of 1000 coulombs at 56 days. These limits will effectively mandate the use of either blended cements or SCM's in such concretes.
Beyond prequalification, the Ontario Ministry of Transportation (MTO) is using coulomb limits based on random coring, for payment in its End-Result Specifications (ERS) for high-performance concrete structures.

3. NEW DEVELOPMENTS

One must decide what environmental exposure and then test methods are relevant for a particular application. Some of the possible tests are listed below:

Penetrability Tests: ionic diffusion, vapor diffusion, rate of absorption, permeability, complex transport such as wick action, and other combinations depending on boundary conditions.

Specific Durability Tests appropriate for the various attack mechanisms and boundary conditions that the structure will be exposed to.

Volume Change Tests for shrinkage (autogenous and drying), thermal changes, and creep.

Concrete Materials

Performance tests for aggregate exist for many properties, although gradation limits maybe seen as prescriptive. There has been a lot of activity at ASTM C01 to develop performance standards for hydraulic cements (ASTM C 1157) but the performance of cement is influenced by other components of the increasingly complex paste fraction of concrete: cement, supplementary cementitious materials (pozzolans and slag), a series of chemical admixtures, together with water. Issues such as workability, set time, rate of strength development are a function of all of these components in combination, and cannot be predicted from performance testing of any one component. Therefore a new ASTM joint C01/C09 subcommittee on cementitious materials-admixture compatibility has been formed in 2005 (from a task group formed in 2003) to try and develop test methods and standard practices for evaluating performance of the total paste fraction of concrete. One of these is a draft standard practice for *Monitoring Cement-Admixture Interactions using Isothermal Calorimetry*.

Hardened Concrete

Concrete with low drying shrinkage properties is becoming more common in specifications. Prequalification of concrete mixtures for low-shrinkage exists in the Australian AS 1379, and has just been introduced as an optional performance requirement for prequalification in CSA A23.1. The CSA test uses ASTM C 157 but with a 7-day wet cure, followed by drying at 50% rh, with a 28-day limit of 0.04%. However, this is not total performances since shrinkage stresses and cracking involve other processes. We also need tests for autogenous shrinkage.; there is a RILEM committee currently working on this. We also need tests for restrained shrinkage cracking potential;

AASHTO has a test method and ASTM has a draft. These properties then need to be combined with early-age thermal and creep properties and a knowledge of geometric properties and ambient conditions to fully predict (using modeling) potential for early-age potential for cracking.

Performance Specifications

The responsibilities of the various parties need to be clearly defined with a performance specification. This has been attempted in the newly issued CSA A23.1-04, the essence of which is shown below. In addition, an Appendix was added to the standard which explains these issues in more detail.

Alternative	The owner shall specify	The contractor shall	The supplier shall
(1) Performance: When the owner requires the concrete supplier to assume responsibility for performance of the concrete as delivered and the contractor to assume responsibility for the concrete in place.	(a) required structural criteria including strength at age; (b) required durability criteria including class of exposure; (c) additional criteria for durability, volume stability, architectural requirements, sustainability, and any additional owner performance, pre-qualification or verification criteria; (d) quality management requirements (e) whether the concrete supplier shall meet certification requirements of concrete industry certification programs; and (f) any other properties they may be required to meet the owner's performance requirements.	(a) work with the supplier to establish the concrete mix properties to meet performance criteria for plastic and hardened concrete, considering the contractor's criteria for construction and placement and the owner's performance criteria; (b) submit documentation demonstrating the owner's pre performance requirements have been met; and (c) prepare and implement a quality control plan to ensure that the owner's performance criteria will be met and submit documentation demonstrating the owner's performance requirements have been met.	(a) certify that the plant, equipment, and all materials to be used in the concrete comply with the requirements of this Standard; (b) certify that the mix design satisfies the requirements of this Standard; (c) certify that production and delivery of concrete will meet the requirements of this Standard; (d) certify that the concrete complies with the performance criteria specified; (e) prepare and implement a quality control plan to ensure that the owner's and contractor's performance requirements will be met if required; (f) provide documentation verifying that the concrete supplier meets industry certification requirements, if specified; and (g) at the request of the owner, submit documentation to the satisfaction of the owner demonstrating that the proposed mix design will achieve the required strength, durability, and performance requirements.
(2) Prescription: When the owner assumes responsibility for the concrete.	(a) mix proportions, including the quantities of any or all materials (admixtures, aggregates, cementing materials, and water) by mass per cubic metre of concrete; (b) the range of air content; (c) the slump range; (d) use of a concrete quality plan, if required; and (e) other requirements.	(a) plan the construction methods based on the owner's mix proportions and parameters; (b) obtain approval from the owner for any deviation from the specified mix design or parameters; and (c) identify to the owner any anticipated problems or deficiencies with the mix parameters related to construction.	(a) provide verification that the plant, equipment, and all materials to be used in the concrete comply with the requirements of this Standard; (b) demonstrate that the concrete complies with the prescriptive criteria as supplied by the owner; and (c) identify to the contractor any anticipated problems or deficiencies with the mix parameters related to construction.

Table 1: Prescriptive v. Performance Specification Responsibilities adapted from CSA A23.1-04.

In addition, a number of highway agencies have adopted end result specifications (ERS) with contractor penalty and bonus clauses for meeting performance requirements, based on tests on random coring. In Ontario, the Ministry of Transport (MTO) requires contractors to pay for random coring and testing for Strength, Air-Void parameters, and Chloride permeability (duplicate cores are taken for check tests by the MTO). Penalty and bonus payments to the contractor are based on both meeting the specifications and on the uniformity of results. In 2005, MTO is introducing prequalification of concretes for one bridge contract based on 0.06% 28-day shrinkage, with initial measurements at 1 day and without additional curing. Virginia is also adopting an ERS, but only based on strength and air void structure.

Related to this, the NRMCA Foundation has recently issued the first phase of a contract to K. Hover, J. Bickley, and R.D. Hooton, with the goal to develop a model performance specification.

In-Place Testing

Owners of concrete structures only care about in-place performance of the structure. One example of what is needed is a test to evaluate the chloride resistance of cover concrete resulting from the concrete quality, the compaction, finishing, and curing. It is difficult to determine whether concrete has been well cured. Relative rates of absorption between surfaces and at some depth below the curing-affected zone could be used. But how can this be used to determine its effects on service life?

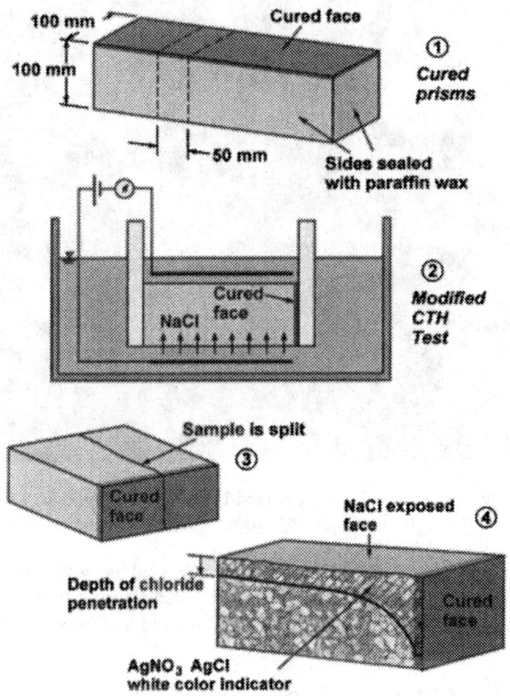

Figure 3: Modified Rapid Migration Test to Determine Curing Effect on Chloride Resistance (Hooton, Geiker, and Bentz, 2002)

A test to measure the chloride diffusion properties of concrete as a function of depth from finished or formed surfaces was proposed by Hooton, Geiker, and Bentz (2002). In this test cores can be taken and saw cut to obtain a prism perpendicular to the surface. The Nordtest NT Build 492 rapid migration test is then run and the chloride penetration profile with depth is visualized with a spray of silver nitrate solution and measured. This is schematically shown for a lab prism in Figure 3, and a typical profile is shown in Figure 4.

Figure 4: Mortar (0.50 w/c) Cured 1-day then Air Cured to 28 days, with Finished Face on Left, Showing Higher Chloride Penetration Near the Surface

Using the non-steady state diffusion coefficient calculated using the NT 492 equation, in Figure 5, diffusion values were calculated at 5mm depth intervals for a 0.35 w/cm concrete with silica fume and slag. These values could then be input to computer models to evaluate the impact of curing on time to corrosion.

Can Tests and Standards Ensure Performance?

Not totally, but they can dramatically improve the situation. If we want to avoid problems, then all concrete contracts need to be treated like high-performance concrete contracts and the different "players" have to communicate and discuss all the performance issues. Get all the players on board early. The owner, the architect, the engineer, the contractors, the concrete suppliers, and inspectors need to work together to achieve the required service life. A key person from each party needs to be a champion for Service Life before the job starts and as construction progresses. Some agencies require contractors to cast a mockup section. For flatwork this could be a test section using the contractor's equipment and crews. In Ontario, first-time contractors using the MTO HPC bridge specification are required to do this. The contractors, the subs, including the finishers, and the suppliers need to be aware of what needs to be done to ensure that the concrete can be delivered, placed, compacted, protected, finished, and cured to achieve the service life objective. Even the person who will be fog misting , or applying other protective measures needs to be there to understand why it is important.

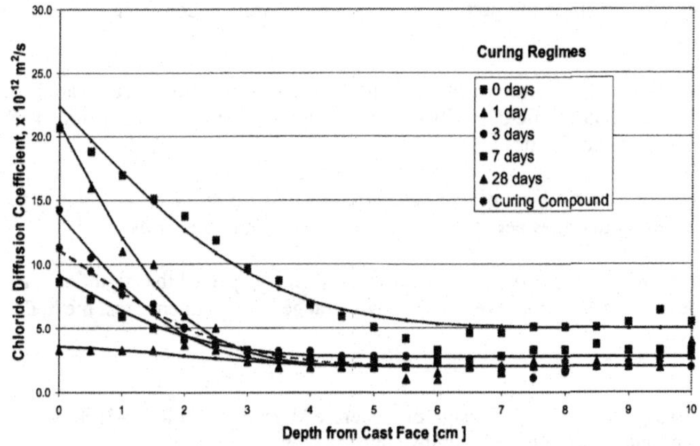

Figure 5: 28-day diffusion values as a function of depth for different curing regimes for 0.35 w/cm concrete with 6% silica fume and 25% slag (base value = 2 x 10^{-12} m^2/s) (unpublished data, My Phuong Ha and R. D. Hooton)

Workability:

- The specified workability must be sufficient for the timing, size, and type of pour as well as the quantity and spacing of reinforcement.
- Specifying stiff concrete can lead to undesirable additions of water to the mixer or during finishing operations.
- The contractor must be able to handle the concrete, so the contractor should set the slump requirements for his/her construction constraints.

Inspection Needs Include:

- Formwork quality and tightness
- Re-bar/PT placement and clear cover checks
- Concrete placement issues
- Vibration equipment and access
- Protection application
- Finishing (e.g., water issues)
- Curing application and timing
- Joint installation

CONCLUSIONS

It is clear that in order to change current prescriptive specifications towards performance specifications, new test methods are needed and improvements are needed to existing test methods, to better address all of the performance issues and provide confidence to owners and specifiers.

We do need better tests that address relevant performance and durability issues, are based on materials science, and relate to field performance. As well, associated specifications need to adopt reasonable performance limits

However, tests cannot cover all conditions, so either testing needs to be conducted for job-specific conditions, or modeling is needed to extrapolate to different scenarios.

With all the current interest in performance specifications, it is expected that significant advance will be made in development or improvement in a wide range of performance test methods.

REFERENCES

Abrams, D. A., 1918 (revised 1925). "Design of Concrete Mixtures", Bulletin #1, Structural Materials Research Laboratory, Lewis Institute, Chicago, 22pp.

CSA A23.1-04, 2004. "Concrete Materials and Methods of Concrete Construction", Canadian Standards Association, Toronto.

Hooton, R. D., Geiker, M. R., and Bentz, E.C., 2002. "Effects of Curing on Chloride Ingress and Implications of Service Life", ACI Materials Journal, Vol. 99, No. 2, pp. 201-206.

Monteiro, P.J.M. and Kurtis, K.E., 2003. "Time to Failure for Concrete Exposed to Severe Sulfate Attack", Cement and Concrete Research, Vol. 33, No. 7, pp. 987-994.

Roumain, J. C., 2002. "Standards and Concrete Durability: An Industrial View" Designing For Concrete Durability, Anna Maria Workshop, Florida.

Wilson, R. and Cleve, A., 1928. " Brief Summary of Tests on the Effect of Sulfate Soils and Waters on Concrete", Report of the Director of Research for the Year ending November 1928, Portland Cement Association, pp. 70-74.

Peter Emmons and Fred Glasser

Arnon Bentur and Jacques Marchand

Duncan Herfort

Juri and Karin Gebauer

Denis Mitchell, Sidney Diamond and Jan Olek

Niels Thaulow, Karen Scrivener and Emery Farkas

Listening to the speaker?

Sidney Diamond and Arnon Bentur

A "nano" discussion

During a presentation

A moment of relaxation

Author Index